Studies in History and Philosophy of Science

Volume 46

More information about this series at http://www.springer.com/series/5671

Tzuchien Tho

Vis Vim Vi: Declinations of Force in Leibniz's Dynamics

 Springer

Tzuchien Tho
Università degli Studi di Milano
Milan, Italy

ISSN 0929-6425   ISSN 2215-1958   (electronic)
Studies in History and Philosophy of Science
ISBN 978-3-319-59053-0   ISBN 978-3-319-59055-4   (eBook)
DOI 10.1007/978-3-319-59055-4

Library of Congress Control Number: 2017940801

Printed on acid-free paper

This Springer imprint is published by Springer Nature
The registered company is Springer International Publishing AG
The registered company address is: Gewerbestrasse 11, 6330 Cham, Switzerland

To Amy Marie

Acknowledgements

The research that went into this work was supported by the Berlin-Brandenburg Academy of Sciences and Humanities, the Max Planck Institute for the History of Science, the Institute for Research in the Humanities (University of Bucharest) and the University of Milan (Statale). It also benefited from a short visiting research period at Princeton University. I must thank Eberhard Knobloch, Vincenzo Di Risi, Dana Jalobeanu, Mihnea Dobre, Stefano Di Bella, Gianfranco Mormino and Daniel Garber for their direct support of this project. O. Bradley Bassler first set me on the Leibnizian path. I would also like to thank the many scholars in the field of early modern (and Leibniz) studies who helped solidify my project: Richard T.W. Arthur, Edward Slowik, David Rabouin, Valérie Debuiche, Emily Grosholz, François Duchesneau, Maarten Van Dyck, Maria Rosa Antognazza, Charles Wolfe, Tobias Cheung, Stephen Howard, Lucian Petrescu, Davide Crippa, Ansgar Lyssy, Erdmann Görg, Kirsten Walsh, J. Colin Mcquillan, Christopher Noble, John Bova, Tal Glazer, Anne-Lise Rey, Lucio Mare, Luca Guzzardi, Rodolfo Garau, Pietro Omodeo, Angela Axworthy, Delphine Bellis, Karin Verelst, Arnaud Pelletier, Ursula Goldenbaum, Lloyd Strickland, Amanda Hicks, and Justin E. H. Smith. Special thanks also go to the two blind reviewers of the manuscript who pointed out some ambiguities and embarrassing mistakes, and to Andrew Ascherl for his invaluable proofreading help.

This work would also have been impossible without the support of my friends. To Pietro Bianchi, Giuseppe Bianco, Olivier Surel, Samo Tomšič, Krisha McCune, Benjamin Bishop, Patricia Goldsworthy, Bruno Besana, Mae Liou, Aleks Zarnitsyn, Etienne Chambaud, Valentina Suma, Eve Richens, Brian Howell, Nate Vacher, Ryan Anderson, Daniel Kugler, and Victor Lao, I owe a debt of gratitude.

Contents

Abbreviations

A   Leibniz, G. W. 1923-. *Sämtliche Schriften und Briefe*. Ed. Academy of Sciences of Berlin. Darmstadt, Leipzig and Berlin: Akademie Verlag.

AG   Leibniz, G. W. 1989. *Philosophical Essays*. Ed. and trans. R. Ariew and D. Garber. Indianapolis: Hackett Publishing.[1]

C   Leibniz, G. W. 1903. *Opuscules et fragments inédits*. Ed. Louis Couturat. Paris: F. Alcan. Reprinted 1988. Hildesheim: Olms.

GM   Leibniz, G. W. 1849–1863. *Mathematische Schriften*. 7 vols. Ed. by C. I. Gerhardt. Berlin and Halle: A. Asher and H. W. Schmidt.

GP   Leibniz, G. W. 1875–1890. *Die Philosophischen Schriften*. 7 Vols. Ed. by C. I. Gerhardt. Berlin: Weidmannsche Buchhandlung. Reprinted 1960–1961. Hildesheim: Olms.[2]

L   Leibniz, G. W. 1969. *Philosophical papers and letters*. 2nd ed. Ed. and trans. Leroy E. Loemker. Dordrecht and Boston: Reidel.

LC   Leibniz, G. W. 2001. *The Labyrinth of the Continuum: Writings on the Continuum Problem, 1672–1686*. Trans. Ed. and Intro. R. T. W. Arthur. New Haven and London: Yale University Press.

LdB   Leibniz, G. W. 2007. *The Leibniz-Des Bosses Correspondence*. Trans. ed. and intro. Brandon C. Look and Donald Rutherford. New Haven and London: Yale University Press.

LdV   Leibniz, G. W. 2013. *The Leibniz-De Volder Correspondence. With Selections from the Correspondence Between Leibniz and Johann Bernoulli*. Trans. Ed. and Intro. by Paul Lodge. New Haven and London: Yale University Press.

---

[1] Abbreviations AG, C, GM are cited by series, volume, and page.

[2] Abbreviations GP, L, LC, LdB, LdV are Cited by volume and page.

Chapter 1
Introduction, Chronology and Historiography

Abstract This introductory chapter of the book lays out the basic motivation and goals of the book. The chapter develops along the lines of outlining the fundamental idea for a systematic interpretation of Leibniz's dynamics project, provides a brief chronology of Leibniz's engagement with the project and explains some historiographical choices made in the periodization of the texts of the dynamics.

1.1 Some Introductory Remarks

The title "Declinations of Force" is an attempt to use grammar to make a number of allusions. This linguistic connotation is not to be taken literally as the subject of the monograph but provides an oblique approach to a difficult and often slippery subject. First, the title hints at the way that articles, nouns, pronouns, and adjectives decline within a grammatical structure. On this point, it should perhaps be noted that the term "*dynamica*" was one of Leibniz's own invention from Latinized Greek and is plural neuter rather than singular feminine like "*mechanica*".[1] In French, Leibniz strangely wavers between the singular feminine "*ma dynamique*" and the plural "*mes dynamiques*".[2] In German, Leibniz appears to be more consistent, with "*unsere dynamica*" (A III 5, 597). This is perhaps an elaborate way to draw attention to the importance of context. Force is not, or is not merely an entity or form. It plays a role within the account of a transforming theory of the causation of motion. This role and how the concept of force fulfils this role is still debated today. Hence, force declines because it only has meaning through its role within a range of other physical concepts and problems of its time that had not yet found stable expression.

At the same time, the specifically Leibnizian definition of force, a concept in stark contrast to the standard Newtonian concept, appears in the history of early modern physics like an archaic grammatical case, replaced in actual use by other terms. The term, energy-work ($1/2mv^2$), is the closest analogue in standard classical mechanics to Leibnizian force (mv^2). In this sense, like the ablative case in modern German, for example, the declination of force alludes to something that was facing decline and obsolescence but also something that played a central role in the process of the stabilization of the "grammar" of classical mechanics, re-inscribed into another form.

---

[1] This point is made by Fichant (1995, 51).

[2] See A III 6, 506 and A III 6, 618.

© Springer International Publishing AG 2017
T. Tho, *Vis Vim Vi: Declinations of Force in Leibniz's Dynamics*, Studies in History and Philosophy of Science 46, DOI 10.1007/978-3-319-59055-4_1

Along with these senses of declination is also modification. The concept of force and the causal role within Leibniz's account of motion also developed and was modified. Our aim here is to understand this declination qua development within the structure of Leibniz's evolving philosophical systems.[3] Leibniz's force thus "declines" insofar as it develops into different ways in which bodies and motion are related. Just as different syntactical forms allow us to parse sentences according to subject-object, direction and possession, the development of the concept of force allows us to parse the different statements of the dynamical theory according to subject-object, direction, possession, and modification. Force, through the years of Leibniz's work on the concept, will appear as entity, relation, quality, and tendency from different perspectives. Force could thus be understood through its various declinations. Tracing the development of the concept of force is a tracing of how one series of variation transforms into another series.

In a more directly metaphysical sense, moving away from a strictly grammatical metaphor, declination also refers to an ancient tradition of treating non-mechanical deviation in physics. In an early text by Leibniz, *De prima materia* (1670–1671), he brings out the powerful Epicurean-Lucretian image of the rain of matter, each portion of matter following a parallel trajectory with respect to each other. Calling on the Epicurean tradition, he claims that, "But that in which there is no variety is not sensed. Similarly: If all matter were to move in one direction, that is, in parallel lines, it would be at rest, and consequently would be nothing" (A VI 4, 279; LC 344). Here, Leibniz notes that only an angular transformation in the direction of a portion of matter could allow for the origin of the cohesion of matter. In the context Leibniz sought to argue for circular plenum motion. Nonetheless, this is precisely what Lucretius called the "*declinatio*" (alternatively, also *inclinatio* and *clinamen*) of the parallel "rain" of atoms (Lucretius 2002, 119). Of course, Leibniz would be very clear that he rejects both the "vulgar" atomism of the ancients as well as its more sophisticated revival through the work of his contemporary, Gassendi, but this provides a wonderful image of the fundamental role played by force within the dynamics. Force, for Leibniz, represents the presence of a non-mechanical principle that grounds all mechanical processes. Unlike Lucretian declination, Leibnizian force is not a declination in the sense of a non-mechanical event in an indeterminate "metaphysical" past that determines the phenomenon of a closed mechanical present. However, like Lucretius, Leibnizian force calls upon a non-mechanical process responsible for generating the principle and forms regulating the shape of the physical world. Force qua cause is actually present in mechanical processes but constitutes non-or infra-mechanical causation insofar as it does not reduce to the efficient or operational interaction of bodies. Force thus declines in the sense that it introduces a non-mechanical factor to every mechanical process. It is, as Leibniz, puts it, an "incorporeal" principle in corporeal phenomena (GP IV 107; L 111).

---

[3] One might also allude here to magnetic "declination," a theme dear to Leibniz's natural scientific and engineering research. Many earlier writings show Leibniz's deep interest in the intersection between magnetism, horology, and navigation (*circa* 1672–1676). This is somewhat outside of the scope of materials I can examine here. Nonetheless, it could be noted here that Leibniz did understand magnetism as a force distinct from gravity and elasticity (pressure) (A VIII 2, 17).

Now, there is a final sense in which we can speak about the dynamics project as a declination of force. Force, in the Leibnizian sense, is the cause of motion as the cause of a structure. What is this structural causation? This is the question that will be taken up in the following chapters. Briefly though, I claim that, as a quantity of conservation, force qua mv^2 (the product of mass and the square of speed)[4] is an invariant that is realized across a series of different values (of the magnitudes of motion) distributed among bodies in a physical system. Here the same cause is expressed through a series of different relavtive locomotive phenomena. Conversely, a variety of different equivalent locomotive phenomena is individuated through a single cause. Hence, force gives rise to a variety, and a variety is founded through a unifying cause. In this sense, dynamical causality is structural since the relation of causation operates between heterogeneous domains.

Force produces phenomenon. Cause and effect, or force and phenomenon, in the dynamics, do not belong to the same domain. Hence, the relation between non-mechanical cause and mechanical effect is related only through a structure which produces this effect. The cause qua force causes effects through the production of a variation of extensional relations within physical systems. Structural cause means that the relation between cause and effect can only be expressed through an organized variation of effects. Analogically, since nouns, pronouns, and adjectives decline through a grammatical structure, a locomotive phenomenon is only given through the declination of force in its distribution in space and time. In this sense, the measure mv^2, the measure of Leibnizian force, should be distinguished from force understood as the cause of motion (not just the quantity conserved in elastic collisions). This measure mv^2 is the invariant governing series of variables expressed by motion. This measure provides an account of how force qua cause translates into phenomena qua effect. Of course this translation is necessary precisely because the relation between cause and effect in the dynamics occur neither between phenomena nor between operational forces. That is to say that Leibnizian force is not causal in the immediately physical sense. It is rather causal through the mediation of a structure governing the appearance of physical motion or locomotive phenomenon. Like a convergent infinite number series with a determinate law but whose terms are never fully given, a physical phenomenon is produced through a causal and mediating structure that is not directly given by the phenomenon itself. Hence, force declines in the sense that dynamical causation cannot be grasped as a physical entity but only through its declination, that is, through the variables defined under the structural organization of the phenomenon of motion.

---

[4] I will use "mass" and "speed" in the more general sense of the terms throughout this book. For "mass" Leibniz uses "*moles*" or "*ratione corporum*" rather consistently and begins to develop a theory of specific gravity as early as 1677 as evidenced by the sophistication of "*De modo pervenendi ad veram corporum analysin*". A theory of mass in the sense of canonical classical mechanics was clearly defined in *Dynamica*, however Leibniz does not always have in mind the scientific definition. For "speed," Leibniz often means velocity and represents linear directions with negative variables. Again, his use of the term from the early period to the late is not consistent, and *celeritas* and *velocitas* are used interchangeably. I use the terms without any specific appeal to their technical definition.

It is this final sense of "declination" that is most important to the interpretive project that follows. The analogy to grammar is most salient in this respect. Primitive force as the formal cause of bodies and motion is often presented by Leibniz to be his answer to rival Cartesian and mechanist philosophers of his age (GM VI 239–240; AG 123). It is supposed to refute the Cartesian or otherwise mechanist thesis of inert bodies that asserts the exclusive causation of physical motion through external, or mechanical-efficient causes, replaced by Leibniz with inherently active bodies. This metaphysical claim of the inherent action of bodies, playing the role of the unextended soul in traditional metaphysical systems, does not, in itself, require any scientific account of motion (Chaps. 2 and 5). On the other hand, a physical theory grounded in the measure mv^2 as the invariant proportion in any motion does not require any metaphysical thesis. It is only in the translation of a metaphysical thesis concerning cause and effect to the scientific account of physical motion that Leibniz's dynamics project becomes legible as a project. Primitive force is entelechy (ἐντελέχεια), an inherent capacity to move (GM VI 236; AG 119). Although Leibniz's use of the term is a heterodox version of the Aristotelian-Scholastic framework from which he takes this term, this use is relevant in that primitive force is an entity that exists only insofar as it allows other predicates to exist in its subject. Hence, primitive force can never be expressed as such. It can only appear as a structure that orders physical phenomena and governs relations between bodies.

Of course, I do not wish to carry this "structuralist" thesis too far in the interpretation of Leibniz. The broader "structuralist" idea can be grasped in a number of texts in Leibniz's corpus. It is not my intention to defend Leibniz as a "structuralist" in the many domains where he has been characterized in this way: (proto-)linguistics, the foundation of mathematics, or the concept of metaphysical harmony.[5] I defend "structural causation" as the concept of causation in the dynamics only in the sense that I have indicated above. That is, force constitutes a cause for motion insofar as force structurally causes the variation in the locomotive phenomena that constitutes the physical world. Broader questions about the "structuralism" of Leibniz's philosophy will not be addressed here.

1.2   Starting Points

The explanation of the title "Declinations of force" in the above stands for an account of the motivations for this book and an indication of its basic interpretive goal. In what follows here, I will provide a brief history of Leibniz's dynamics

---

[5] Several of these "structuralist" themes are directly relevant to Leibniz's long-term projects on linguistics. The various works on *grammatica rationalis, lingua rationalis, charateristica verbalis* around 1678–1679 demonstrate a structuralist or proto-structuralist approach to lingistics. The project sought to reduce all grammatical variations in Latin to a small number of elements and then use this to map other European languages. In an encyclopedia project of 1679, *Consilium de encyclopaedia nova conscribenda methodo inventoria*, Leibniz places this project of grammar in the first order of subjects (A VI, 4, 343). See also A VI, 4, n. 21, 22, 24, 35, 36.

project. The aim here is to provide a general chronological map that will orient the reader to the central texts that will guide my interpretation. It will also provide some historiographical reasons for the organization of these texts.

Let us start in the middle. The immediate historiographical problem facing Leibniz's dynamics is its start and its end. Here, I have followed many other interpreters in tracing the start of the dynamics project at 1676, with the text *De arcanis motu*. Along with these interpreters, I also place the "end" of the dynamics at 1700–1701. This span of 25 years is not obvious. For Leibniz himself, the dynamics only started in 1689 when the term was created between the writing of the *Phoranomus sive de Potentia et Legibus naturae* in the summer of 1689 and the *Dynamica: De potentia et legibus naturae corporeae*, a few months after in the fall of 1689. This is roughly the "middle" of the period that I am proposing (1676–1701). The term *"dynamica,"* dynamics, a term familiar to us from physics classes, was of Leibniz's own invention from the Latinized Greek δύναμις in 1689 (A III 4, 483–486).[6] However, we can see that since the dynamics was understood as a science of force, this concept in Leibniz's writings certainly predated the term *"dynamica"* by several decades. The term force, *vis*, and the related term power, *potentia*, was already a significant aspect of Leibniz's work before 1689. We see that Leibniz was already using this term in the mid-1670s. Of course these terms were inherited by Leibniz through ongoing debates about the cause of motion starting with the important forebears of the "scientific revolution" that had been raging in Europe since the end of the sixteenth century.

Between inheritance, development, and maturation, where does the dynamics project start? I defend the start of the dynamics project in 1676 due to the reorganization of Leibniz's physical theory from the perspective of causation. This is met with some complications. Certainly Leibniz is well-known as a mathematician and metaphysician, a reputation that was foregrounded by his immediate reception in the "Leibnizianism" of the Wolffian school but also remains for the most part unchanged in his reception in the nineteenth and twentieth centuries. For this reason, Leibniz's physical theory is largely attached to either the metaphysics of a "dynamicism" or to the mathematics of the calculus. We cannot address the complex history of this development here, but what this most occludes is the fact that Leibniz's earliest published writings concerned physical theory. In the early period of Leibniz's work, the 1660s to the early 1670s, we see Leibniz focused on a project to harmonize an Aristotelian world of formal, final, material, and efficient causes with the contemporary philosophical trend towards a mechanistic interpretation of the physical world that held efficient cause as the causal concept in which all other causes could be dissolved (A II 1, 17–19). This early period is characterized by scholars as displaying the underlying ecumenicalism that will pervade Leibniz's thought until the end of his life. What is difficult to explain in this is the choice of physical theory as a point of departure for his manifold ecumenicalism, ranging from theology and metaphysics to law and ethics. What is especially important for us is Leibniz's deeply rooted concern for physical theory, a domain from which other theological and metaphysical problems could be addressed. This is a commitment that can be traced to the late 1660s.

---

[6] See Fichant (2016, 11–41, 12–16).

If the dynamics project is founded on the concern for the causes of motion, then perhaps this is a project that goes right back to the very start of Leibniz's public career. In the two treatises of 1671, *Theoria motus concreti* (*Hypothesis physica nova*), received by the Royal Society, and the *Theoria motus abstracti*, received by the Académie royale des sciences, Leibniz was not so much concerned with the problem of the causes of motion as with the form or shape of motion. We will address these problems with a little more care in the present book, but it suffices to say that what separates the early Leibnizian treatment of motion from his later work is the concern for causality. In the early work, Leibniz treated cause as a subset of the problem of the geometrical constitution of motion from its parts. Some of the insights from this period runs through Leibniz's work until the end, but the young thinker would face enormous intellectual challenges as he leaves Mainz for Paris in 1672. The cliché that Paris would change everything for Leibniz is certainly true. Of course, the city itself and his official diplomatic mission had little to do with his transformation. It was rather that in Paris Leibniz was initiated into the networks of philosophical and scientific communication radiating from London, Geneva, Amsterdam, Leiden, Rome, Hamburg, and other centers of publishing and intellectual activity.

The four years in Paris, mentored by Huygens, provided Leibniz with the conditions to produce significant breakthroughs in mathematics. Part of this mathematical maturation will be reflected in his rejection, along mathematical lines, of the earlier physical treatises in 1671. This is not our focus here. Along with the shedding of mathematical naiveté, Leibniz was also hard at work in his study of the mechanical treatises of the day. This period reflects Leibniz's in-depth study of Galileo, Huygens, Fabri, Wallis, Mariotte, and others. By the time Leibniz left Paris in 1676, he was conversant in contemporary mechanical problems and even drafted treatises aimed at solving problems in the engineering of machines, the philosophical foundations of optics, on the alluring phenomenon of magnetism and raging debates about astronomy.

In this period of the late 1670s, Leibniz was prepared to reframe mechanics based on his understanding of the physical and mathematical problems, and some deeply-held metaphysical commitments. Despite his disagreements with Cartesians on the inertness of bodies, his idea of the period remained largely Cartesian in style and content. This idea of "reforming" mechanics remained the explicit aim of much of Leibniz's physical theory until the arrival of the "dynamics" properly speaking.[7] However, among a number of different writings on the subject from 1676–1678, *De arcanis motus* stands out as the text which would provide the foundations of the abandon of the mere "reform" of mechanics to the "new science" of the causes of motion. Indeed, he refers to his realization as "secrets".

Now *De arcanis motus* remains a text that still aimed at contributing to a reform of Cartesian mechanics. It lays out rather systematically Leibniz first reflections on how to properly investigate the causes of motion. Leibniz saw it appropriate to call his principle an "axiom" on par with Euclidean axioms of whole and part. "Just as the first axiom of geometry is that the whole is equal to all its parts, hence the first

---

[7] See Fichant (1994, 59).

mechanical axiom is that the full cause and entire effect has the same power [*potentia*]" (A VIII 2, 59). There are two additional features of this text that lead us to place it so centrally. The first is the highlighting that the study of motion based on its causes needs to be based on a capacity for "future" effect, thereby separating cause and effect through time. The second is the outline of a mathematical theory for comparing equipollent forces that would provide an absolute evaluation of the magnitude of force or power between different motions. We will provide a closer examination of these themes in due time (Chaps. 2 and 5), but it is for these reasons that we place this text at the start of the dynamics project.

Another text of the same period rivals *De arcanis motus* as the founding document of the dynamics project. This is *De corporum concursu*, Leibniz's attempt to write a physical treatise based on the insights developed in *De arcanis motus*. What is central about this text is his first adoption of the measure of mv² as the conserved quantity of force. To be clear, Leibniz did not discover this conservation quantity. Huygens had already argued for this in the case of collision in his contribution to the 1669 *Philosophical Transactions*, a text that Leibniz read even before his close relationship with the former in Paris. Mariotte, it should also be mentioned, also argued for this conservation quantity in experimental terms in 1673, under the influence of Huygens, in his *Traitté de la percussion ou les chocq des corps* (Mariotte 1673). Leibniz studied this text quite in depth and had a correspondence with the celebrated savant. To make matters even more complicated, when Leibniz began to write *De corporum concursu*, he still had in mind the Cartesian quantity of motion mv (with direction). It was only half-way through the composition of the text that Leibniz saw the centrality and absoluteness of the quantity mv². Halfway through the composition of this text Leibniz returns to the first page in order to cross out his initial principles related to the Cartesian quantity of motion stating, "This does not follow from our system" (Leibniz 1994, 71).

Another complementary aspect of the eventual dynamics project also solidified in this period of the mid-1670s. While the phenomenon of extended motion has always been held by Leibniz to be derivative to some more fundamental ontological basis, Leibniz's reasons for this wavered between a logical one and an ontological one. Logically speaking a body is either in one place or another place, *tertium non datur*. Since motion is change of place, it is neither in one place nor another, hence it cannot be predicated of a body. Ontologically speaking, the state of rest and motion of a body is relative to these states with respect to other bodies. Thus the property of motion is in itself indeterminate. During this period, in texts like *Principia mechanica* (c. 1676), Leibniz arrived at a way to address all these issues through the borrowing of the Keplerian notion of the equivalence of hypotheses. We shall address this in more depth in the present book. Briefly, this principle states that a single locomotive event can be described across different assignments of relative rest and motion among relatively situated bodies. It is the same to say that a body *approaches* a resting body at V speed as to say that a resting body is *approached by* a body moving at V speed. This solidifies a manner of approaching the measurement of motion as a relational property that obeys structural invariances in cases like motion in a non-accelerated inertial frame.

This approach to the relational account of motion is also not unique. He follows a line of important influences including Galileo and Huygens. With regard to Kepler, Leibniz's adoption of the principle of the equivalence of hypotheses is complex since, even as he draws from this influence, he seems to have the beginnings of the modern notion of inertia, against the Keplerian notion of the "inclination to rest" [*inclinatio ad quietem*]. Nonetheless, with this concept, Leibniz does put into play an important aspect of the theory of force that is not a direct outcome of equipollence of cause and effect. This will significantly contribute to the maturation of the dynamics.

If we look backward from the mature dynamics of 1689 and early 1690s, we can take all of these concepts as foundational to it in some way. As I have pointed out above however a number of these terms and concepts existed before 1676, and a number of these ideas would only become related or solidified after 1676. Placing the origin of the dynamics more than a decade before the actual coining of the term means that we forego any "smoking gun" for the origin of the dynamics project. What is salient however is that in 1678 and subsequent years, Leibniz engages in a number of attempts to compose a book-length treatise on "physics" based on the conjunction of ideas first organized in the 1676 *De arcanis motus*. What we have from these attempts are outlines and prefaces of a book or books that were never written. Indeed, we can see *De corporum concursu* as part of these frequent attempts. However once the ideas of the equipollence of cause and effect, the equivalence of hypotheses, the theory of future effect, and the conservation quantity of mv² were put together, Leibniz was certainly well on his way to the dynamics. For this reason, we take 1676 as the start of the dynamics project.

Now, despite the seriousness with which Leibniz engaged the development of a physical theory in the late 1670s, there would be almost a decade that separates these writings and the next burst of texts in the mid-1680s. There would then be a few more years after that before the dynamics would emerge properly speaking. Given the fact that problems of force, body, and motion never left the many significant writing projects of the time, it is incorrect to say that the dynamics project slumbered. What is significant, however, is that in 1686, a key date in the development of Leibniz's systematic metaphysics, the completion of the *Discours de métaphysique* and the start of his prolonged correspondence with Arnauld, Leibniz publishes the *Brevis demonstratio erroris memorabilis Cartesii*. This would be the start of what I call the second phase of Leibniz's dynamics project. With the basic elements in place and their initial relations worked out into a theory, Leibniz began to write publicly about its more controversial implications. The *Brevis demonstratio* is a polemical text that states very little of the positive theory behind his refutation of the reigning Cartesian physics. However, it shows that Leibniz had matured enough in his thinking about the dynamics to go on a public offensive against the Cartesians of his day. In the responses to this text in journals and correspondences, Leibniz was also able to show his fellow savants that a complex and systematic theory lay behind such polemical essays.

If the 1676–1678 texts formed the start of the first phase of the dynamics, the second phase starts around 1686 and ends around the writing of the *Dynamica* in 1689. What is important to the second phase is the development of the explicit metaphysical aspects of the dynamics project. The principle of equipollence and the alternative conservation quantity mv^2 was put forward as a rejection of a strictly mechanistic world. The attempt to argue against the mechanistic closure of efficient causes was already the intention of Leibniz's work since the late 1660s, but the development of the dynamics contributed to how this intention bloomed into the systematic metaphysics of substantial form which was the hallmark of Leibnizian metaphysics of the 1680s and early 1690s. Of course the principle of equipollence and the measure mv^2 were developed within the context of physical theory. It is difficult to see why the mere replacement of mv with mv^2 as the law of motion and collision would require a different metaphysical picture. The point, however, is to appreciate that Leibniz saw, in his nascent dynamics project, doctrines convergent with his systematic metaphysics. Rather than providing necessary or sufficient conditions, the dynamics project provided a number of different ways to bring fundamental metaphysical principles together with the coherence and harmony of physical reality.

An important aspect of this "metaphysical" second phase of the dynamics project was the emphasis on final causes. Most importantly, the equipollence of cause and effect relies on the concept of future effect. The conservation quantity is thus expressed through the empirical reality of efficient causes but acts to realize a final cause that cannot be reduced to mere efficient cause. This is the most immediate sense of why a replacement of a conservation quantity would provide evidence of a metaphysics of causes incompatible with the Carteisian one. We shall address this later, but we note here that this second phase is important precisely because it allowed Leibniz to definitively break with the idea of a reform of Cartesian mechanics. A transformed theory of causation meant first and foremost, that a physics must be grounded in a different metaphysics.

Since a large portion of the commentary on Leibniz's dynamics has focused on the themes most salient to this second phase of the dynamics project, I have chosen to restrict my remarks on many of these themes here.[8] There is a wealth of excellent commentaries from to Leibniz's earliest reception in the Wolffian school to recent interpretations. It is also my intention to suggest that, although I have little to criticize in the work of these commentators, it is through this strong association of the dynamics with the metaphysics of the 1680s that some of the development of the dynamics leading up to this point, as well as some of the significant shifts after this point, has been occluded. Some recent authors have provided new avenues that guided my research. Among the most significant of these interpreters, Fichant, Robinet, and Duchesneau have argued that the dynamics takes a significant departure in 1689 towards a different physical theory. My work here is an attempt to extend the indications made by these interpreters in the move towards treating the significance of the dynamics beyond the lens of the physical-metaphysical convergence in the 1680s. Of course, there is no way to minimize the importance of this phase in Leibniz's dynamics project, but Leibniz's work after 1689 constitutes the most important aspect of the current project.

---

[8] For the standard and contemporary reception of this period of the dynamics (the "second phase"), I have in mind the interpretations of M. Gueroult (1934), M. Fichant (1995 and 2004), D. Garber (1985, 2009) and, G. Gale (1988) and G. Brown (1984).

From October 1687 to June 1690, Leibniz embarked on a long voyage through south Germany, Austria, and Italy. In this extremely productive voyage, Leibniz composed, among other works, two extended physical texts. These were the *Phoranomus* and the *Dynamica* mentioned above. Both of these texts reflected his discussions with the many savants he visited during his voyage and especially his extended association with the Accademia fisicomatematica in Rome. Without going into details here, it is important to emphasize that there were a number of different significant events that coincided with this voyage. First, Leibniz claims to have read Newton's *Philosophiae Naturalis Principia Mathematica* during his voyage. This is probably not true (Bertoloni Meli 1993, 8). Regardless of when Leibniz actually read Newton's *Principia*, the writings of 1689 certainly reflect a shift in the theoretical terrain. Leibniz's physical theory was no longer aimed at responding to the dichotomy between late Scholasticism and Cartesianism. A third position weighed in heavily. It should be noted that Newton represented a challenge in debates over astronomy. The Newtonian theory of gravitation and its account of the inverse square law directly provided the challenge for Leibniz to explain the transradial movement of planets on the basis of the motion of the subtle matter of the plenum. More foundational disputes concerning the cause of motion were never part of the debates between Leibniz and Newton or the Newtonians (with the exception of a few additional episodes on this issue between Leibniz and De Volder in friendly correspondence). Nonetheless, the account of planetary gravitation was a hotly debated issue before Newton's identification of attraction and mass. In this context, Leibniz provided an account of the inverse square law based on a synthesis of earlier plenum theories. In 1689, Leibniz composed and published a number of significant but controversial treatises on planetary motion.[9]

It should be cautioned that these texts are often confused with the dynamics for the reason that they are closely related to problems of motion and gravitation. However, as we will see, one of the aims of the mature dynamics was to dissociate gravitation, the cause of which is the relation between planetary motion and the motion of the plenum, from force as the cause of motion as such. In any case, the effect of Newton's rise as the dominant figure in physics made it insufficient for Leibniz merely to face Cartesian opponents. Through his writings on planetary motion and in the emergence of the dynamics, it was Newton and the Newtonians who provided the new intellectual context in Leibniz's work in the late 1680s.

Second, in the previous year Leibniz had resumed his correspondence with Huygens, which had faced a complete freeze due to the latter's unfavorable reception of the former's geometrical projects in 1680. The theme of this resumption of communication was over the subject of isochronal and tautochronal curves and more broadly the laws of motion. Eight years after their last correspondence (a correspondence gap from 1680 to 1688), Leibniz reengages with Huygens on the problem of the Cartesian attacks on his *Brevis demonstratio* (A III 4, 368–371). Huygens does not respond for close to a year. When the correspondence regains speed, Leibniz and Huygens would begin establishing a response to the new Newtonian opponents of

---

[9] Most significant is *Tentamen de motuum coelestium causis* (GM VIII 144–187).

this question of the laws of motion. Whereas Leibniz and Huygens saw their contributions to physics as circulating within a scientific milieu flanked by late scholastic and Cartesian opponents, neither thinker saw their work as orthodox, and neither embraced Newtonianism. Both Leibniz and Huygens held to expanded versions of Galilean relationism about motion. Here, although Huygens' views concerning the Cartesian plenum was complex, on the one hand presenting a version of plenum theory in the 1690 *Discours de la cause de la pesanteur,* and on the other hand, in its appendix, defending an atomistic theory of void and perfectly indeformable bodies, the new Newtonian opponents created a common front for Leibniz, Huygens, and the Cartesians against what they saw as the unacceptable position of attraction at a distance.[10]

The two points above provide the context for the emergence of the dynamics properly speaking. But the flowering of the dynamics project itself occurred precisely between the writing of the *Phoranomus* and the *Dynamica.* Both of these texts, a dialogue and a treatise respectively, composed one after the other with some overlap in the summer and fall of 1689, were ambitious works that sought to provide comprehensive expositions of Leibniz's mature physical ideas. Neither text, as we have them now, were completed. The *Dynamica*'s incompleteness has been subject to much speculation. Due to its near completion and clear evidence of active revision over a number of years, it is less clear why this treatise was never completed. However, the incompletion of the *Phoranomus* seems to follow much more straightforwardly. The *Phoranomus* was the last attempt to provide a comprehensive account of Leibniz's physical theory without the concept of action. The crucial bridge between the two texts is that the concept of action, which accounts for the inherent activity of a body to bring about motion, was sketched out in an unfinished argument in the *Phoranomus.* This concept was unnamed in the former text. Through Leibniz's work on the *Dynamica,* not only was the Graecism "*dynamica*" firmly installed, the concept of action (*actio*) was also formally defined as the product of a body's state of motion (speed) and the magnitude of the displacement of a mass already travelled. We shall examine the concept of action in more detail across the chapters of this book but for the moment, we can appreciate how the rise of the dynamics proper coincides with the new terminology of action.

Alongside the definition of a new theoretical grounds (action) from which to base the dynamics, the *Dynamica* also provides a broad synthesis of Leibniz's mature physical theory from the theory of specific gravity (specific weight) to a sophisticated understanding of the composition of motion. Although many of these ideas are significant for Leibniz's maturation in the domain of physics, what is most important for us here is the transformation of the concept of cause. It is hence precisely in this passage from the *Phoranomus* to the *Dynamica* that a new concept of cause, action, would definitively provide the account of how force causes motion. It is thus this shift in the dynamics project that will guide our retrospective look at the work leading up to this point and the work in the dynamics that followed.

Although much of the dynamics was already fixed in 1689, another decade of work on this project followed. The last phase of Leibniz's dynamics project, the

---

[10] Huygens 1944, 451–486. See Mormino 1996, 76–78 and 2011, 697–705.

work from 1690 to 1700, is perhaps the most significant for its reception. Since the *Dynamica* was neither completed nor published, texts like *Specimen dynamicum*, the two texts entitled *Essay de dynamique* of 1692 and 1700–1701 stand out as the only available expressions (excepting correspondences) of this large and developed intellectual project that had come to (partial) fruition in 1689. What is most significant about this period comes from the fact that the *Dynamica* had rendered the idea of an inherent action of bodies concrete in the concept of force as the cause of motion. As such, the dynamics becomes both a fully developed physical theory about motion as well as a robust metaphysical theory of the action of substances. In texts of the post-1689 period, the results of the dynamics were employed as direct references for the constitution of substance. In texts like *De prima philosophia emendatione et de Notione Substantiae*, *Specimen dynamicum*, and *De ipsa natura*, the notion of force was given the role of expressing substantial form in the metaphysical hylomorphism that Leibniz had sought to present in different ways since his youth. More precisely, force plays the role of constituting inherent force in bodies set apart from the extensional or phenomenal aspect of bodies.

In this later period, Leibniz will increasingly use the distinction between primitive and derivative force to describe a primary and formal metaphysical role played by primitive force (inherent substantial action), and the measurable extended effects of force (derivative) as modalities or modifications of force as it is distributed or "diffused" (*diffundi*) in space and time (LdV 304–305). In this way, the metaphysical foundation provided by primitive force, equated with the entelechy (ἐντελέχεια) of substance, plays the role of grounding extended and measurable physical phenomena, neatly placing force outside of the dimension of extended things even as it participates in generating the world of extended phenomena.

Now, the vast majority of commentators tend to place this metaphysical distinction between primitive and derivative force at the center of the dynamics. Leibniz already speaks of a primitive force of resisting and acting *(vis resistendi primitiva* and *vis agendi primitiva)* attached to matter and substantial form respectively in the mid-1680s (A VI 4, 1507–8, LC 285–7). However, the distinction plays no role in the main texts of the dynamics like the *Phoranomus* and *Dynamica*. This distinction only becomes a key point of reference only after the mid-1690s in texts like *Specimen dynamicum* and the following correspondences with De Volder, Wolff and others. Of course, interpreters follow Leibniz himself in the texts of the 1690s in using this distinction to decipher the importance of these dynamical concepts. However, it is important to clarify that this distinction is central only to the metaphysical exposition of the dynamics. Hence, while there is no surprise that Leibniz's metaphysical ideas are often infused in his scientific projects and vice-versa, the attempt to understand the fundamental problems of the dynamics is occluded by placing metaphysics in the foreground. My fundamental quarrel here is that when the dynamics is seen through the perspective of its metaphysical exposition, the comprehension and evaluation of the dynamics become attached to desiderata proper to Leibniz's larger metaphysical systems as they evolve. Instead, what is sought in the following chapters is the examination of the dynamics from its stated intent. That is, to understand Leibniz's development of the concept of force as a "new science" which unfolds a theory of physical causation. From this perspective, the many sub-distinctions of force, primitive/derivative, partial/total, passive/active, dead/living, will be considered under the main perspective of causation. In the same vein, while commentators since Leibniz's immediate reception in seventeenth and eighteenth century tend to use the term *vis viva* (living force) to refer in general to Leibnizian force, I will speak about *vis viva* as a counterpart to *vis mortua* (dead force). These terms indicate certain roles played by force in physical systems but they refer to a fundamental causal concept.

The last phase of the dynamics project was also the most public and discussed through important correspondences notably with his friend and supporter Johann Bernoulli, the brilliant young Cartesian-turned-Newtonian Burchard De Volder and Jakob Hermann (a student of Jacob Bernoulli), who synthesized Newtonian and Leibnizian mathematical methods in his physical treatise *Phoronomia, sive De viribus et motibus corporum solidorum et fluidorum*. These correspondences will also be crucial in understanding how the dynamics evolved beyond the text of the *Dynamica* and Leibniz's exposition of its ideas to a large and diverse audience.[11]

Another aspect of the last phase of the dynamics (1689–1701) also coincides with a shift in metaphysics. Although there is no consensus on the precise shape of this shift, the monadic metaphysics of the period after 1695 moved Leibniz closer to an idealism that placed ultimate reality in a world of simple monadic unities imbued with appetite and perception. In this metaphysical picture, physical reality is expressed as phenomena, harmoniously coordinated perceptions shared among monads. There is dispute about how consistently we can interpret this monadic metaphysics to the previous metaphysics of physical reality made up of aggregates of form-matter substances. According to a conventional way of understanding this shift, whether merely terminological or deeply conceptual, the metaphysical aspects of the dynamics seems to have been endangered by the reduction of physical reality to the phenomenal states of monadic perception. I will argue that this apparent danger can be avoided by understanding the dynamics according to the interpretation of force as structural cause. Hence, the dynamics is not only compatible with a number of different metaphysical systems but is also grounded in a metaphysical aspect quite independent of larger ontological concerns.

Before Leibniz's death in 1716, he engaged in a heated and polemical correspondence with Samuel Clarke. Fifteen years after the end of the dynamics project, Leibniz continued to refer to the concepts that he had developed since 1676. What is curious from the perspective of the dynamics is that, although concepts like force were introduced in the correspondences with Clarke, Leibniz seems to have left the metaphysical and systematic ideas of the dynamics project out of this debate. In some sense, he seemed to have seen his differences with the Newtonian camp as focused on theological issues concerning the will and intellect of the divine creator. Although the debated problem of the absoluteness of motion will be important for understanding a crucial aspect of the dynamics, Leibniz's response to Clarke as a

---

[11] The correspondence with Hermann would last until the year of Leibniz's death in 1716 (GM IV 260–416).

whole pushes us to see that the dynamics as the science of force qua the cause of motion cannot stand in for Leibniz's mature physical theory as such. Theology, astronomy, chemistry, and many other domains contribute equally in Leibniz's constantly adjusting concepts for a systematic physical theory. The dynamics constitutes merely one aspect of this, albeit an important one.

Although the project was never completed, Leibniz continued to refer to the dynamics in key writings until his death. As we have mentioned, there is speculation on the reasons for its incompleteness. I can provide no good reasons for this. What will be argued instead is a systematic interpretation of the dynamics project from the perspective of its central aim: an account of the cause of motion.

1.3   A Quick Outline of the Book

Although this monograph is historically oriented in tracing the chronological development and context of Leibniz dynamics, it will be organized according to the conceptual elements that constitutes this project as a whole. After this introduction, I will provide, in the second chapter, a summary defense of the central interpretation of structural causation from a developmental perspective. In the third, fourth, and fifth chapters, I will present the three fundamental elements of the dynamics. In the third chapter, I will examine the role of the principle of the equivalence of hypotheses for the project. In the fourth chapter, I will trace the concept of continuity of motion in its often ambiguous evolution. In the fifth chapter, I will treat the equipollence of cause and effect in its role as the central "axiomatic" foundation for the dynamics. These three chapters will demonstrate why an interpretation according to structural causation makes the most sense out of the documents and key concepts of the dynamics project. In the sixth and final chapter, I will argue for the compatibility of the dynamics with the monadic metaphysics of Leibniz's later years.

Before I end this introduction, I should note that the image of "Leibniz" that emerges here may be foreign to many seasoned scholars. Across the reception of Leibniz, since the days of Wolffian "Leibnizianism," Leibniz appears in kaleidoscopic aspects of each different philosophical turn (vitalist, pan-psychic, logicist, computational, organicist, phenomenological, etc.). There is no attempt here to address, in any global way, what kind of thinker Leibniz was. The aim is to address the dynamics project itself, a project of roughly two-and-a-half decades where he struggled with the questions of the inherence of motion in bodies, its measure and its relation to the fundamental laws of nature crafted by the divine.

This monograph will undoubtedly leave open many more questions than it can answer. Leibniz's thought from 1676 to 1701 involves complex breaks, missteps, and detours, and even more complicated continuities. By proposing a "structural" understanding of force as causation, I have only proposed to trace the outlines of a labyrinthine architecture.

Chapter 2
What Is Structural Causation?

Abstract The second chapter of this book presents the central argument of the book in a synoptic form. It proposes a first approach to an important interpretive lens for understanding Leibniz's dynamics project as presenting a theory of causation that is structural in nature. In the preceding introduction, we examined the chronological development of the dynamics and emphasized the growth of the dynamics project from the notion of Leibnizian *vis* as a structural *property*, the property of a physical or mechanical *system*. Now we take a step further. Since the dynamics aimed at developing a science of the cause of motion, the proper object of the dynamics is the nature of causation rather than the mere properties of motion. A structural form of causation is proposed here in order to render explicit the central concept of Leibniz's dynamics.

2.1 Introduction

What is structural causation? In the simplest terms, structural causation is the form of causation that correlates one cause to a multiple set of correlated effects.[1] For the dynamics, the most direct way to appreciate the concept of *vis* is through the gap between un-extended *vis* and extended motion. We take *vis* to be defined as the cause of extended motion but this motion, through the equivalence of hypotheses, is a multiple set of related effects. We will examine the equivalence of hypotheses more directly in the following chapter (Chap. 3) but for the moment, we emphasize that the deployment of force into motion is realized in a set of different phenomena, based on different hypotheses of rest and motion assigned to the body or bodies under consideration. To say that body a is moving toward body b, at rest, is equivalent to saying that body a is at rest and body b is moving towards it. As long as the *vis* of the system is conserved, both hypotheses correspond to the same cause.

Structural causation is thus based on a causal relationship between *vis* and phenomenon (or *phenomena* since it is a one-many relation). This means not only that

---

[1] The terms "structure" and "structural causation" are used in various other contexts in contemporary philosophy of science and political philosophy. My use of these terms does not stem from a dialogue with these existing forms. It only shares with them the idea of privileging systematic or structural principles as causal rather than attributing causality to powers or tendencies located in individual entities. It is ultimately motivated by my other research in analytical mechanics.

© Springer International Publishing AG 2017

T. Tho, *Vis Vim Vi: Declinations of Force in Leibniz's Dynamics*, Studies in History and Philosophy of Science 46, DOI 10.1007/978-3-319-59055-4_2

body-to-body mechanical or efficient causation is cast as derivative to the primary concept of *vis* and subsumed under a more fundamental form of causation, but also that the concept of causation from the Aristotelian tradition, where some potential is actualized through motion, is, if not rejected, then heavily revised (See Chap. 5). The causal relationship here is an operation between a quasi-metaphysical and infra-phenomenal principle, *vis*, and the phenomenal and extended reality that it engenders.

The most convenient analogy that we can appeal to here, Leibniz's own, is that of the Apollonian cone (A VI 4, 1369–1371). In geometrical analysis of the Apollonian cone, the cone itself remains static. However, if a two-dimensional plane transects the cone, we produce the circle, the ellipse, the parabola, hyperbole, and all the continuous morphisms between them (Fig. 2.1).

These different two dimensional "effects" are produced by the motion of a plane through the higher-dimensioned "structure" of the cone. These various geometrical entities (circle, ellipse, parabola, etc.) are also (insofar as they are continuous) point-to-point isomorphic. The one-to-many correlation between cause and phenomenal effect can then be analogically understood as the multiple maps between a higher-dimensional object (*vis*) to a set of internally (group) ordered lower dimensional objects. In other words, dynamical causation is structural because it is the causation that bridges one level of reality (*vis*) and another level (extended phenomena).

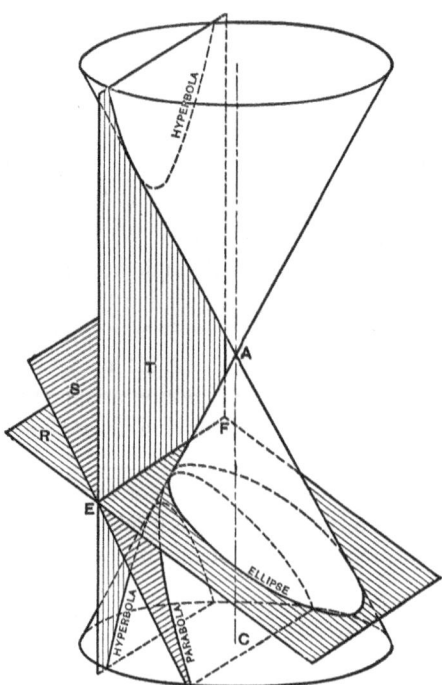

Fig. 2.1 Planar sections of the Apollonian cone (Figure taken from Educational Technology Clearinghouse. 2011. http://etc.usf.edu/clipart/32400/32427/_32427.htm. Accessed on 20 April 2016)

This description of structural causality may, for the instant, rely too much on the reader's post-Kantian intuitions about the noumenon-phenomenon distinction. This later distinction equally plays on the notion that phenomena is epistemically limited in its capacity to present reality in the absolute sense. The key difference between Leibniz and Kant here is that the phenomenal and the real are not separated in an epistemically fundamental way. Leibniz's dynamics project requires that the mind grasps the causes of motion and the account of phenomena according to "metaphysical considerations" beyond geometry and mechanical relations (GP II 13). Hence even if the term "phenomena" undoubtably rings with a Kantian tone, physical phenomena, for Leibniz, are expressions of the modifications of *vis* in space and time. In *Dynamique et Métaphysique Leibniziennes*, Gueroult leveled a strong criticism of this Kantian prejudice against Couturat and Cassirer: "For Leibniz, the foundation of the laws of the sensible is the intelligible itself.... [T]he laws of the first is not simply the double of the intelligible world but is its translation" (Gueroult 1934, 188). The interpretation of *vis* according to structural causality follows Gueroult in the aspect of deepening the Leibnizian concept of phenomenon as modification and translation while being wary of possible Kantian prejudices. Of course, there are further profound and worthy topics to explore in the connections between Leibniz and Kant, but we cannot treat these points in the current work.

The relation between *vis* and phenomena is part of Leibniz's own terminology in the dynamics. Our aim shall be to treat this relation through the notion of structural causality as it occurs within Leibniz's dynamics. The main aspect of this structural causality is the mathematical form through which this passage between cause and effect is realized. *Vis* then, understood not only as the cause of motion, but measured according to the conservation quantity mv^2, provides the concrete bridge between cause and effect. This allows us to understand the quantity not only as the measure of force or power in a mechanical system but also as the operation that renders this form of causation concrete in the dynamics. In what follows in this chapter, we will make the argument for a theory of structural causality through the methodological developments in the measurement of quantity of force mv^2.

The examination of Leibniz's methodology, its aims, its successes, and its gaps reveal the concept of causation at work. In the second section, following this introduction, I lay out the relation between the concept of the conservation of *vis* and the methodology based on *potentia* as structural property of mechanical systems that was needed to realize such a conservation. The aim here is to address the theoretical foundations of the dynamics through the different methodological elements that came together during Leibniz's quarrel with Cartesian mechanics in the 1670s to the 1680s. In the third section, I introduce the concept of *actio*, which was central to the establishment of Leibniz's dynamics, property speaking, in 1689. The aim in this second section is to show how the concept of *actio* provides a new way to understand the nature of quantity conservation and the nature of *vis*. The generality and actuality achieved through the concept of *actio* allowed the dynamics a proper footing, moving the concept of causation away from its initial framework based on *potentia* by incorporating an account of the temporality of the deployment of *vis*. Finally, in the fourth section, I examine Leibniz's final contribution to the dynamics

project, the *Essay de dynamique* (c. 1700–1701). Here we find the mature methodology of Leibniz's dynamics at work. In this examination, I make the case for understanding the concept of causation at work as a structural one. With these three
sections, I aim to explain the concept of structural causation through the developmental context of the dynamics. Much of the details, to be covered in later chapters,
will be glossed over. We aim here to establish the lens through which the dynamics
will be examined further on in the book.

2.2 Equipollence of Full Effect and Entire Cause: The Nature of the Conserved Quantity mv²

The quantity mv^2 served as a guiding insight of the dynamics project. Although we
have dated the start of the dynamics project to 1676, the texts of 1678 introduced the
quantity in a stable form. This holds until the end of the dynamics project in 1701. The
quantity comes most directly out of Leibniz's study of Huygens and Mariotte, and its
rise to central importance is well documented in Leibniz's *De corporum concursu* of
1678. To understand Leibnizian *vis* it is thus important to understand its measurement
and the methodological problems Leibniz encountered. We will then get a clearer
picture of how this measure was able to secure the path to the dynamics.

Before doing this we should clarify one important issue concerning terminology.
In the seventeenth century, the concept of "force" emerges out of the mechanical
tradition as a family of different terms and intuitive representations: pressure, pro-
pensity, moment, power, impetus, and the like.[2] The innovations of Kepler, Galileo,
and Torricelli in this tradition formed the immediate background of Leibniz's pre-
decessors or near-contemporaries (Descartes, Mersenne, Wallis, Fabri, Huygens,
etc.).[3] The terminology was far from settled even within Leibniz's own work.

---

[2] We point here to terms from a mechanical tradition that certainly includes the ancient Greeks
schools of Archimedes and Pseudo-Aristotle but there is also a metaphysical tradition that informs
Leibniz's process of developing the dynamics. The Aristotelian tradition of metaphysical terms
like δύναμις, ἐνέργεια and ἐντελέχεια resonate with key aspects of Leibniz's dynamics. Certainly,
the profound articulation of the distinction between ἐνέργεια and ἐντελέχεια, Aristotle's own
neologisms central to the core of his theory of actuality, were not part of Leibniz's metaphysics.
Further, there is very little evidence of Leibniz's direct engagement with the Aristotelian corpus on
these issues. Given Leibniz's complex relation with the "Schoolmen", the background intent to
restore concepts like substantial form and final causes bows to the Aristotelian tradition while lack-
ing in the faithful adoption of traditional arguments or doctrines. While Leibniz frequently invokes
ἐντελέχεια, written in Greek to highlight the appeal to antiquity, he does not stop to expand on the
term but rather signals the shared educational background of his learned audience. Moreover, like
the nineteenth century employment of the terms "energy" from ἐνέργεια (Thomas Young), the
obvious appeal to δύναμις in the coining of "dynamica" is also far from being situated within an
Aristotelian context. Here, late Scholasticism is much more pertinent as a source of influence. In
particular, the "equipollence of entire effect and full cause", treated in more depth in the fifth chapter, is a term that stems directly from the Scholastic tradition.

[3] Kepler's *Astronomia nova* looms large in these conversations about force, and this text can be
seen as an initial starting place for post-Renaissance thinkers who saw force in connection with the

Nonetheless, we must distinguish Leibnizian *vis* from what we now understand as "force." The term taught to us in textbooks, the vector expression ($\vec{F} = m\vec{a}$), refers, thanks to Euler's formalization, to the Newtonian concept which served as the cornerstone of the formation of classical mechanics. This is the formalization of Newton's second law of motion, "The change of motion is proportional to the motive force impressed; and is made in the direction of the right line which that force is impressed" (Newton 1972, 13). Newtonian force here corresponds to the concept of the change of motion (or momentum), one of the many candidates for the concept of force. This is not Leibniz's *vis*.[4] Bracketing its historical development, we can say that Leibnizian *vis* corresponds more closely to our concept of energy-work expressed by the quantity $1/2mv^2$. This functions as a conservation quantity and is a property of a mechanical system. In this sense, it is not obviously causal.

Bearing this distinction of Newtonian force and Leibnizian *vis* in mind, let us then proceed with clarifying the concept of *vis* from the perspective of the problem of measurement.

During the Paris period (1672–1676), alongside his numerous mathematical projects including the method of the infinitesimal calculus, Leibniz was immersed in the study of mechanics. Wallis, Huygens, Pardies, and Mariotte were the main focus of study.[5] Now, Leibniz had already been deeply influenced by the publications of Wren, Wallis, and Huygens in the 1669 *Philosophical Transactions* concerning the rectification of the Cartesian laws of motion through the examination of elastic and inelastic collision, but such an influence was not seriously taken up until the late Parisian period (1675). Indeed, the main thrust of Leibniz's earlier work on the problem of corporeal motion, from the twin *Theoria motus abstracti* and *Hypothesis physica nova* (*Theoria motus concreti*) until the *Pacidius Philalethes* dialogue (1676), was phoronomy. That is, the problem concerned how to map physical motion onto geometrical relations and properties (continuity, uniformity, dimension, etc.). The phoronomic project sought to treat motion through its compo-

---

ἐνέργεια concept of Aristotelian and Scholastic provenance and its *inherence* in bodies. See the explanation of this concept in Johannes Kepler (III 1990, 240–242). For Galileo and Torricelli, we can refer to the general influence of the "Galilean school". The impact of this school is too obvious to point to one or two texts. However, the kind of discussion of force through collision problems can be found in the short appendix to *Della scienza mecanica: e delle vtilita che si traggono dagl'instromenti di quella* and in the missing "sixth day" of *Dialogo sopra i due massimi sistemi del mondo*. Marin Mersenne produced a French translation *Les mecaniques de Galilée* in 1634. A whole generation of Cartesians and other savants would certainly have used this translation to promote and debate these force concepts. Of course, Torricelli's own work, apart from Galileo, was also known to Leibniz, filtered through the previous generation of savants like Wallis and Fabri. For what concerns the famous Torricelli principle in *De motu gravium* (from the book I of the treatise *Opera Geometrica*) that Leibniz will draw on for the dynamics, we see an extended series of notes in his reading of Honoratus Fabri's *Physica* (Book VI, 85) circa 1670–167 (A VIII 2, 489). John Wallis also comments directly on this in the section on hydrostatics in the *Mechanica* (Pars III, Cap. XIV, 14) (Wallis 1670, 742–743). See also Leibniz's 1674–1675 notes on the discussion of Torricelli and Dati in Wallis' *Mechanica* (A VIII 2107).

[4] From this point on, I will consistently refer to Leibnizian "force" as "*vis*" and Newtonian force as "force" unless otherwise noted.

[5] See these studies in A VIII 2, 57–140; A VIII 2, 263–264.

sition. Hence instead of conservation quantities or the conceptualization of *vis*, Leibniz was more concerned with the uniformity and continuity of motion, what Leibniz would later call the "labyrinth of the continuum" (LdV 333). The problem that Leibniz confronts in the late Paris period was a new stage in his work. Much like the results presented in the 1669 *Philosophical Transactions*, the key issue of the time concerned the determination of the laws of motion from the conservation of certain quantities. On the one hand, we have Wallis and Wren, who argued for the rectified Cartesian quantity mv (the product of mass and directional speed instead of simply mass and magnitude of speed m|v|). On the other hand, we have Huygens and Mariotte arguing for the conservation of mv² (product of mass and the square of speed). Leibniz heavily favored the latter as we shall see.

We should note here that there is nothing mathematically contradictory about the conservation of both mv and mv². Given a system of rectilinear collision for the bodies A and B, before and after collision:

$$m_A v_A^2 + m_B v_B^2 = m_A v_A'^2 + m_B v_B'^2$$

implies:

$$m_A \left(v_A - v_A' \right) \left(v_A + v_A' \right) = -m_B \left(v_B - v_B' \right) \left(v_B + v_B' \right)$$

and vice-versa.

This follows by simple algebra. Nonetheless, this algebraic form was never put forth by Huygens or Marriotte. In the 1692 *Essai de dynamique*, Leibniz did eventually provide for the algebraic derivation of the conservation of force from relative speeds (velocity) and conservation of momentum (Leibniz's quantity of progress) (GM VI 227–228). In addition, we could also mention that Huygens sought to demonstrate his conservation principle from a thought experiment while Mariotte, drawing from Huygens, took the conservation principle as the second of the four "suppositions" which he then verifies through a series of empirical experiments (Mariotte 5–6). Huygens' 1669 publication in the *Journal de sçavans* formulates the principle in the following way: "The sum of products made by the size of each hard body, multiplied by the square of speed, is always the same before and after their collision" (Huygens 1929, 73; 1977, 590). Following this formulation, Mariotte notes, "And reciprocally, the bodies that fall from different heights from their own weight on the same horizontal surface, meets this surface with different speeds, the square of one to the other, as [proportional] to their heights" (Mariotte, 5). We thus have a proportion, h∝v², the height of free fall (h) is quadratically proportional to their terminal velocities (v) at the base of fall. This proportion h∝v² is what is explicitly taken up by Leibniz during his period of mechanical study.

This conception becomes fully integrated into Leibniz's work in the 1678 *De corporum concursu*, his first attempt at a full mechanical treatise after the Paris period. The text itself was never completed and hence bears the mark of many revisions and missteps. Based on the earlier parts of the treatise, we can see that Leibniz was operating on the intention of reforming Cartesian mechanics based his new studies. The principle of conservation at the start of the text was based on Wallis' rectification (Leibniz 1994, 71). More than halfway through, in a section drawing heavily on Mariotte's experiments, Leibniz writes in a scholium that,

I now see where the error is to be found. The force in bodies should not be measured [*aestimanda est*] from speed and the size of bodies but from the height from which it falls. Hence the heights from which bodies fall are as [a proportion of] the square roots of the speeds in question. [...] Thus generally, the *vires* are in a ratio composed of the simple product of the bodies and the square of the speeds. (Leibniz 1994, 134)

Now, although the manuscript of *De corporum concursu* bears the traces of results pursued through the earlier usage of the conservation of mv and the later usage of the conservation of mv^2, it is clear that Leibniz remains committed to some of the results from this earlier period. The difference concerns generalizability. Leibniz holds that, although Wallis' quantity (understood as *impetus*) is conserved in some cases, it does not hold in general. This is then the reason for why Leibniz holds mv^2, rather than mv; both conservation principles, to be the measure of *vis*.

In order to grasp the generality of mv^2, we should briefly look at his mentor Huygens' argument for such a conservation quantity. Huygens had largely solved the problem of elastic collisions by 1656 at the young age of 27. However, only a summary of this work saw publication in the 1669 *Philosophical Transactions* and the *Journal des Sçavans*. These summary texts covered the basics of his methodology and results. The systematic treatise, *De motu corporum ex percussione*, was to be published posthumously in 1703. One might speculate on the depth of Leibniz's knowledge of the details, but judging from correspondences, it appears that the two had discussed this theme in Paris (A III 6, 131). From the perspective of Leibniz's argument strategy, it appears that he quite readily adopted Huygens' core methodology. Central to this inheritance was Huygens' brilliant use and adaptation of the concept of the center of gravity or center of mass frame.

The center of gravity is well known from statics in the Archimedean and (Pseudo-)Aristotelian mechanical tradition and was developed further by Galileo and his disciple Torricelli. This concept was adapted to motion and colliding bodies by making an analogy between the center of mass, say, the masses and distance to the fulcrum on a lever, to the speeds and masses of two bodies in collision and rebounding with respect to their point of rebounding.

For brevity's sake, we postpone a fuller examination of Huygens' treatise to the next chapter (Chap. 3). In short, Huygens' argument in *De motu corporum ex percussione* is that if we employ an experimental set up where the velocity of colliding bodies could be converted into a vertical motion, the center of gravity presented by the maximal heights of these motion cannot be raised. This is an explicit employment of Torricelli's principle.[6] In turn, when the vertical (falling) speed is converted

---

[6] We usually speak of Torricelli's principle (theorem or law) in terms of the hydrodynamic principle, a sub-case of Bernoulli's principle, where the speed of water flowing out of a spout of a container from a certain height is the same as the speed attained by a drop of water if it fell from that height. Although this law is intimately related to the question of the conservation of energy-work, it is not what we are referring to here. We refer here and in all later references to Torricelli's principle concerning the center of gravity. It states that, "Two descending bodies joined together can

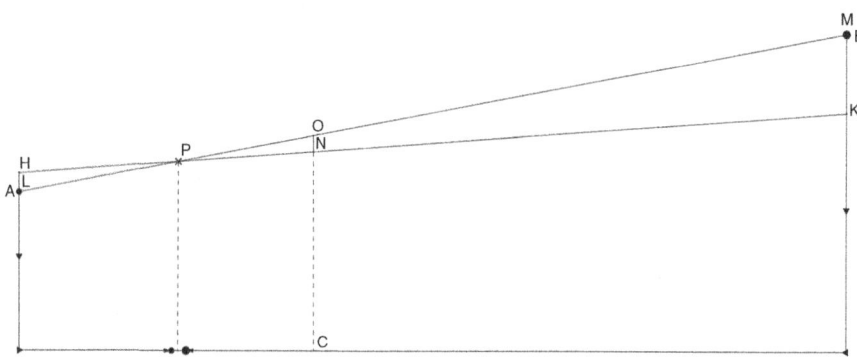

Fig. 2.2 Huygens' experiment for conservation under collision using center of gravity (Figure recreated from Huygens 1929, 59)

into horizontal velocities, the resulting impact will conserve not only momentum but also the capacity to climb back up to respective heights whose proportions maintain their center of gravity. Or rather, the center of gravity cannot be raised (Fig. 2.2).

From this, Huygens reasons that:

$$m_A v_A^2 + m_B v_B^2 = m_A v_A'^2 + m_B v_B'^2$$

The square of velocities is conserved within elastic collisions. As we mentioned earlier, classical momentum is also conserved by implication. What results from this demonstration, outside of the elegant symmetry of the conservation of mv^2 and mv, is the universalization of the concept of Galilean "relativity" or invariance governed by symmetry. This extension of the Galilean principle is the crucial concept inherited by Leibniz through the Huygensian method.

What is crucial to notice here is the use of the Galilean law of falling bodies to satisfy the relation between a principle borrowed from statics and the symmetry of collision. The quadratic relation between speed and height establishes the exponential factor in the conservation quantity mv^2 in the symmetry in collision. The experiment setup that grants Huygens this argument is very difficult to recreate. Given the *reductio* form of Huygens' argumentation, we can take it to be a thought experiment. It was rather Mariotte, based on Huygens' arguments, who found the means to translate such a setup to an empirical experimentation. The idea was to take a double pendulum with bodies of varying masses and heights. Once dropped, the masses meet and rebound, reaching certain heights (Fig. 2.3).

The experiment lacks the property of freefall (a pendulum is constrained motion) but allows us to realize the same conservation quantity. Both approaches to the con-

---

not move by themselves unless their common center of gravity descends... But when the mechanism is configured in such a way that its center of gravity cannot descend further, the body will certainly remain at rest in its final configuration. Furthermore, the body would move in vain because it would be moving in a horizontal line without any downward tendency" (Torricelli 1644, 99). For our purposes, it states that a machine cannot raise its center of gravity and thus solidifies the claim that there can be no perpetual motion machine. See Dugas 1988, 261–263.

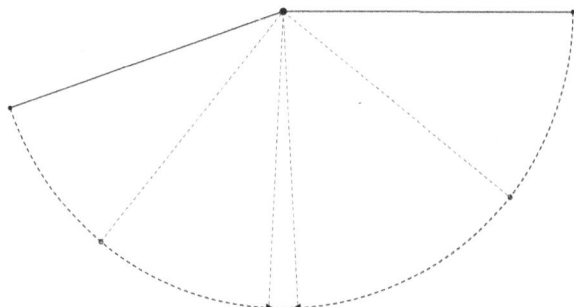

Fig. 2.3 Colliding pendulum experiment demonstrating conservation in Mariotte and Leibniz

servation principle are reproduced in *De corporum concursu* (Leibniz 1994, 131–135; 152–158).

This important inheritance provides Leibniz with the cornerstone doctrine of his eventual dynamics, the conservation of mv^2 qua *vis*. But the generality of this quantity and its method of measurement was perhaps more important to Leibniz than the quantity itself. This Huygensian methodology not only demonstrates the conservation of *vis* in unconstrained, natural, motion (inertial motion) but also applies to constrained, violent motion (in machines).[7] It is this generality of mv^2 that determines how Leibniz will begin to use this conservation principle from 1678 onwards. It is also the reason why mv^2, rather than the Cartesian quantity of motion mv, achieves the measure of *vis*.

One way of understanding the generality of this approach was through the axiomatic foundations laid out by Leibniz for future work in his mechanical research. From his mechanical studies in the 1670s, Leibniz had adopted Huygens' and Mariotte's view of the proportionality between height and speed $h \propto v^2$. A parallel development however is required to fully establish something beyond the laws of motion and render the quantity mv^2 the measure of an entity *vis*. This occurs also in the late 1670s and we find its clearest articulation in the 1676 work *De arcanis motus*. Here Leibniz would set down the fundamental (axiomatic) principle of entire effect and full cause, "Just as the first axiom of geometry is that the whole is equal to all its parts, hence the first mechanical axiom is that the full cause and entire effect has the same power [*potentia*]" (A VIII 2, 59). With echoes of the late scholastic tradition represented in Francisco Suárez's *Disputatio metaphysicae* and Eustachio a Sancto Paulo's *Summa philosophica*, the equipollence (or equivalence) of full cause and entire effect plays a major role in the interpretation of the status of the quantification of the notion of cause.[8] Similar notions of full cause and entire

---

[7] The distinction between "natural" and violent" motion stems from the Aristotelian distinction of the natural motion of bodies by affinity and the motion of bodies due to mechanical constraints. This will be discussed later in this chapter when the distinction becomes contextually salient.

[8] Cf. Suárez, *Disputationes Metaphysicae*, Disputatio XVIII, "Si autem Deus omnia efficit, interrogo rursus an immediate et sufficienti virtute, an mediate tantum et insufficienti virtute.... Si autem primum verum est., superflua est. omnis alia efficientia, quia una causa sufficiens et efficax satis est. ad effectum" (Suárez 2015). Eustachio a Sancto Paolo, *Summa Philosophica* VI 61,

effect also played a heavy role in Hobbes' *De corpore* which was the key influence on Leibniz's early physics (Hobbes, 100–101). Leibniz used the terms "*potentia*" and "*vis*" throughout the dynamics rather interchangeably (Hobbes used the terms *causa* and *potentia* as terms that respectively denote the cause of past motion and the virtuality of the future tendency of a motion), so we should understand the principle of the equipollence of full cause and entire effect as the way in which the quantity mv^2 is the measure of the full cause: *vis* qua *potentia*.[9] The entire effect, in turn, is the full range of motion that it (*potentia* or *vis*) produces. This constrains us to interpret the proportion $h\propto v^2$, regulating the laws of motion, as the measure of a certain quantity. The conservation of mv^2 would thus go from the operation of translating one quantity (height) into another (velocity) within a closed mechanical system to the translation of cause to effect for that system. This principle is recognized by Leibniz as a metaphysical one in *De arcanis motus*, and in a contemporaneous note on 2 December 1676, it is associated with the ontological meaning of the principle of sufficient reason. Here, since nothing is without cause, any effect is ontologically given by its cause *qua* reason (A VI 3, 584; Leibniz 1994, 278).[10] It is this heavy metaphysical (ontological) thesis that is implied when Leibniz places the equipollence of cause and effect in the role of the founding axiom of mechanics not only in *De arcanis motus* but also in a number of other programmatic sketches of the period like the fragment entitled *Tria axiomata primaria* (A VI 3, 427) and *Principia mecanica* (A VI 3, 111; Leibniz 2013b, 116). The internal mathematical organization of the mechanics and its empirical demonstrations thus requires a foundation drawn from metaphysics. Hence this foundation establishes an entity, *vis* (or *potentia*), which is measured by effects which are not homogenous with it.

This dual metaphysical and scientific development solidified in the theory of *vis* comes during the 1680s. The key development is that *vis*, or *potentia*, moves from being one of a series of concepts describing a reformed (Cartesian) mechanics to being the central concept describing mechanical causality. This is the period where Leibniz launches a number of polemical texts against the Cartesians in the form of the 1685 *Brevis demonstratio erroris memorabilis Cartesii* and repeated in the 1686 *Discours de Métaphysique* (articles xvii-xviii). These landmark texts highlight not only the conservation quantity mv^2 but also the generality of its method of measurement. These arguments of the 1680s are based on the measurement of mv^2 from the

---

"Quinta in totalem et partialem; quo spectat alia divisio insufficientem et non sufficientem; illa dicitur quae sola in suo ordine et genere totum effectum producit" (Eustachio 1614, 61). Evidence of Leibniz's familiarity with Suarez is abundant, but key among these is his collection of notes from 1663–1664 on Daniel Stahl's Compendium metaphysicum where Suarez is collected (A VI 1 20–41). Evidence of Leibniz's knowledge of Eustachio a Sancto Paolo, an important figure for Descartes, is less abundant. Leibniz famously makes explicit reference to Eustachius in the 2 February 1706 letter to Des Bosses (LdB 8).

[9] This interchangeability is evident from earlier texts like *De arcanis motus* (1676), where *potentia* is actually privileged over *vis* in the terminology. This is maintained at least up to the *Phoranomus seu de potentia et legibus naturae* in 1689, where we find interchanged uses between *potentia* and *vis* from one paragraph to the next. This is clarified by Robinet in his annotations to *Phoranomus* (Leibniz 1991, 526–527). See also Hobbes, 100.

[10] See Leibniz 1994, 278.

perspective of the capacity to do work, a magnitude measured by the physical effect of a closed system. We understand this as the conservation of the quantity of work more systematically in the concept of energy-work. A theory of causation thus emerges, one that concretizes the vision provided by *De arcanis motus*, the definition of power as the translation of full cause and entire effect. We should examine how this works.

The mood of the central texts of the 1680s was polemical. The aim was a refutation of the Cartesian theory of body and motion by undoing one of its central doctrines, the conservation of the quantity of motion mv. Now in a certain sense, Leibniz's opponent was not Descartes but rather the Cartesians, and his treatment of the quantity mv, though critical, was generous insofar as it already took into account its initial imprecisions that were addressed by nearer contemporaries like Wallis and Wren. Nonetheless, it is clear that Leibniz and Descartes were arguing for different things and attempting to isolate different phenomena. The only real term under question was the universality of the Cartesian conservation quantity, or the quantity that God conserves in the world (*Deo in mundo conservari*) (GM VI 117).

Leibniz's argument unfolds around a possible experiment. He takes two bodies, A and B, with inverse proportions of mass (1 and 4 pounds respectively) and final height (4 and 1 feet respectively). By raising A up 4 feet and B up one foot, the same amount of work (*quanta opus*) will be achieved. The elevated bodies A and B are then let to free fall, and at the base of their fall, using an appeal to Galileo's law of falling bodies, the height (proportional to the duration of fall) will produce a terminal velocity for A that is four times that of B. This conclusion shows that the quantity of work is not conserved by the quantity of motion in freefall. We can see that Leibniz has at least established enough in this case to refute the universality of the conservation of the Cartesian quantity of motion.

$$w = \text{mass} \cdot \text{unit of height, or, } w = m \cdot h,$$
$$w_A = 1 \cdot 4 = 4$$
$$w_B = 4 \cdot 1 = 4$$

And if we calculate for the Cartesian quantity of motion $= m \cdot v$ (mass·velocity) and assume through an analogue of Galileo's law that v (at the base of fall) $= \sqrt{(2 \cdot h)}$

$$v_A = \sqrt{(2 \cdot 4)} = \sqrt{8}$$
$$v_B = \sqrt{(2 \cdot 1)} = \sqrt{2}$$

Hence, for mv,

$$mv_A = 1 \cdot \sqrt{8} = \sqrt{8} = 2\sqrt{2}; \text{or the quantity of motion of A}$$
$$mv_B = 4 \cdot \sqrt{2} = 4\sqrt{2} = 4\sqrt{2}; \text{or the quantity of motion of B}$$
$$\text{and thus } mv(A)/mv(B) = \tfrac{1}{2}$$

We also note that the same example, calculated for mv² results in the following:

$$mv^2_A = 1 \cdot 8 = 8$$
$$mv^2_B = 4 \cdot 2 = 8$$
$$\text{and thus } mv^2_A / mv^2_B = 1$$

In the same text, however, Leibniz wishes to establish a stronger claim, based on the proportion h∝v², that we can find the right conservation through the quantity mv². Since the elevated positions of A and B contain the same amount of power, the translation of this power into the effect of extended motion should then also be the same. In addition, Leibniz takes this to mean that if this same experiment were performed on a pendulum of two bodies, the *vis* of the one body would be capable of raising the second body to its designated height and vice-versa. This is, of course, nothing but the equipollence of full cause and entire effect at work (Fig. 2.4).

Scholars disagree about the exact nature of what Leibniz aimed to accomplish in the *Brevis demonstratio*. G. Brown has argued that the aim was not to refute the Cartesian notion of the conservation of the "quantity of motion" in the world but rather to separate conservation from the Cartesian quantity and introduce the concept of *vis motrix* (the force of motion) as a superior candidate (Brown 1984, 122–137).[11] Regardless of how Leibniz presented such a concept and its demonstration in the mid-1680s, our point here is that Leibniz will not deviate from this basic paradigm for the presentation of the conservation of *vis* in his later writings on the dynamics. Through this, Leibniz would also provide a transparent methodology such as we have been tracing.

Let us then underline a few points in our discussion of the relation between the measure of the quantity mv², the concept of *vis,* and its governing principle, the equipollence between full cause and entire effect.

What Leibniz inherits from the tradition and his near contemporaries was a concept of conservation that determined the laws of motion. When Leibniz seized upon the principle of the equipollence between full cause and entire effect, the ambiguous concept of *vis* achieved the status of an entity and an associated measure based on the translation of quantities between cause and effect. This is the conception of *vis* as the power (*potentia*) of a mechanical system. From this, Leibniz was able to argue for the generality of this quantity of *vis* across different physical situations from constrained motions in machines and freefall and rectilinear collisions.

The crucial nuance to grasp here is that Leibniz's breakthrough in the late 1670s is the solidification of the concept of *vis* and its methodology of measurement more than the universality of the quantity mv² itself. Leibniz was fully open to the idea that a different world could have a different quantity resulting from this methodology (Leibniz 1994, 134). Nonetheless, without his struggle to find the appropriate quantity, this concept would have remained abstract, more metaphysics than physics. In return, the metaphysical aspect of this development produced an appropriate entity for which this quantity could be a measurement.

---

[11] See also Robert Sleigh, Jr. (1990) and Paul Lodge (1997).

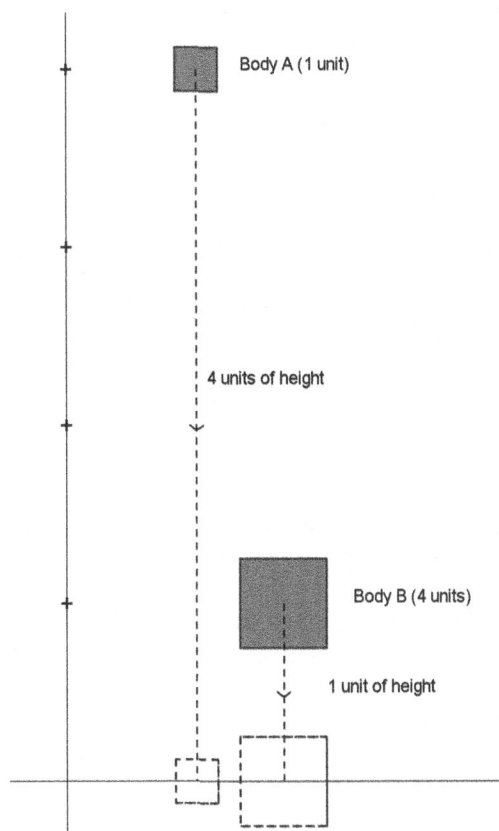

Fig. 2.4 Leibniz's "refutation" of Cartesian force

In what follows, we will continue examining the transformation of the concept of *vis* into structural causation precisely through the transformation of the methodology of its measurement.

2.3   *Potentia, Actio*, and the Origins of the *Dynamica*

The concept of *vis* and its surrounding arguments seem to have been stabilized in the 1680s. However, this was still almost a decade before 1689 when the dynamics itself would be explicitly named. What happens during these years is precisely the maturation of the concept of *vis*, or *potentia*, into the systematic foundation of the dynamics. The maturation of the concept would then allow Leibniz to declare a specific demonstrative science based on *vis*.

The methodology of measurement based on the equipollence of cause and effect had a lingering problem that was not addressed until the late 1680s. According to the principle, the power of full cause is expressed by entire effect. But this "entire effect" takes time. Of course, if we take the equipollence of cause and effect as such, this means that the power of a mechanical system is expressed in the achievement of the full or entire extensional effect of the motion(s) in that system. But if we take the motion of the system *in media res*, part of that motion will be *yet* to be achieved. Future effect is virtual rather than actual as a function of time. Since future effect is not actual, force or power is divided within itself between one factor of causation that is already expended and another factor that will be realized. From this perspective the full conservation of *vis* qua *potentia* could only hold when time is not considered.

The *Brevis demonstratio* was published in March 1686, and in September 1686 the powerful community of Cartesians had already provided a response. In this year the Abbé de Catelan and Malebranche responded to Leibniz's challenge.[12] The Abbé de Catelan, publishing in the *Nouvelles de la république des lettres*, pointed exactly to the blind spot of Leibniz's refutation. Catelan argued that, since there is no account of time in Leibniz's imagined experiment in the *Brevis demonstratio*, one could not establish the equality of "force" in the two bodies A and B. In Catelan's view, the time (and speed) required to raise the two bodies were not accounted for. Hence, it was an error to assign the same quantity of "force" to the two bodies. The two cases are thus inappropriately compared. Catelan's point is incorrect but coherent enough to mount a retort.

The conservation of momentum is indeed conserved in any elastic collision. But to account for this conservation, we would require a different experimental case. For a system of a pendulum with two masses, the demonstration of the conservation of momentum would require a kinematic account of the distribution of speed in the raising of the bodies after the collision of a two body pendulum. Since the properties of the fall of the pendulum would be proportional to its ascent, in such a case a Leibnizian conservation of work would be obvious and conserved. However, a conservation of momentum would account for the change of momentum, the acceleration, required to bring those masses to their maximum speed at the base of their rotational motion and then the final (maximum) respective heights of each body after their ascent. More than this, in other experimental settings the shape of fall and ascent are also relevant for the transformation of speed. The question of the time (qua acceleration) required to raise the two bodies to their respective heights was thus not irrelevant.

However Leibniz's response to Catelan largely ignored the function of time in acceleration. In his response to Catelan through a correspondence mediated by Bayle, Leibniz stuck with the axis of his previous argument. Since the quantity of work is accurately accounted for by the square of velocity, $h \propto v^2$, the temporal account of velocity was not relevant to his refutation of the Cartesian quantity of motion. Leibniz notes, correctly, that the speed of ascent and descent is dependent on its curve. Whether it is freefall, incline, or curvilinear, the speed of a descending or ascending body will depend on the shape of the path of fall. However, this is

---

[12] For Catelan's response, see GP III 40–42. For Leibniz's response to Malebranche, see GP III, 51–55. The *Brevis demonstratio* would be attacked by Denis Papin later in 1689 (Papin 1689, 183–189).

independent of the conservation of the quantity of mv^2 because Leibniz was only interested in establishing the fact that, in a pendulum for instance, a maximal velocity is reached rather than *how* this speed evolves in the path of motion. This is certainly true, but it remains unaddressed in the texts of the late 1680s. Leibniz would eventually have to face the challenge of *why* this was the case.

In 1689, Leibniz nearly completed two major manuscripts within a couple months of each other. The first was the *Phoranomus seu de potentia et legibus naturae,* written in the summer of 1689 while in Rome. The second was *Dynamica de potentia et legibus naturae corporeae,* written in the fall of 1689 on the way back to Hannover (though Florence) from Rome. This second manuscript was Leibniz's attempt to provide a treatise for the "new science based on *vis,*" the dynamics. Indeed, it is around this time that Leibniz coins the graecism "*dynamica*" and begins to use it as a title for his physical program.

As Duchesneau carefully noticed, the difference between the *Phoranomus* and the *Dynamica* is the emergence of the term *actio* (Duchesneau 1998, 77–109).[13] Indeed, this term appears immediately in the *Specimen praeliminare* of the *Dynamica* as the fourth demonstration, whereas it had not been part of the terminology up to this point. It is important to note, however, that the argument surrounding the concept of action appears in the *Phoranomus.* The two texts are conjoined and distinguished by this conspicuous argument that appears in the *Phoranomus* as one among many arguments and in the *Dynamica* as a grounding argument for the entire project.

Although the title of the *Dynamica* is often referred to as "*Dynamica de potentia*" ("Dynamics of power"), as Fichant argues, it should really be referred to as *Dynamica: de potentia et legibus naturae corporeae,* "Dynamics: On power and the laws of the natural bodies" (Fichant 1995, 50). *Vis* and *potentia* receive a new configuration through the introduction of the concept of *actio.* It may be the case that the earlier *Phoranomus* was abandoned precisely because of the emergence of a new foundational concept. I leave this speculation (though a compelling one) aside in order to simply examine this new development in Leibniz's mechanics: the establishment of the dynamics.

The theory of *actio* attempts to deliver a response to the problem of time that Leibniz had brushed off a few years earlier. It approaches the problem of time from a dissection of the mv^2 quantity by separating out the two factors of velocity. One of the velocity factors will be understood as the effect produced by a motion in time. The other factor will be defined as the velocity of the body at the time of measurement. Hence, *actio* will be defined as the product of mass, distance traversed, and velocity over time. This term taken over time arrives at the same quantity $mv^2 = a/t$. Leibniz will call the product of mass and distances traversed the formal effect (ms). He will then call the product of formal effect and the velocity *actio* (msv).[14]

Leibniz's presentation of the concept in the *Dynamica* proceeds through syllogistic form:

---

[13] See also Duchesneau 1994, 147–262.

[14] Elsewhere Leibniz calls this "pure effect" [*effectum purum*] (GM IV 379).

The action bringing about the double [effect] in a single [unit of] time is twice the action of
  bringing about the double [effect] in double the time.
The action bringing about the double [effect] in double the time is double the action bringing about the single in a single [unit of] time.
Therefore the action bringing about the double [effect] in a single time is four times the
  action bringing about a single in a single time. (GM VI 291–292)

We will not remark on the syllogistic presentation of the concept here but only note
that Leibniz argues through the principle of transitivity. Since the first action brings
about twice the effect of the second action, and the second action brings about twice
the effect of the third action, the first action brings about four times the effect of the
third action.

If we take the measure of action as divided between two factors, the formal
effect, ms (mass and distance traversed) and velocity, then the quantity of action for
the three different measures of actions (in time) are in a ratio of one to two and two
to four.

For *actio* (a) as the product of formal effect (m·s) and velocity (v) in time:

$$a = msv$$
$$a_1 = 2ms \cdot 2v = 4msv$$
$$a_2 = 2ms \cdot v = 2msv$$
$$a_3 = msv = msv$$

This explains the proportions given in the demonstration:

$$\frac{a_1}{4} = \frac{a_2}{2} = a_3$$

And we can also explain the connection between the quantity mv^2 and *actio*. The
application of *actio* above was treated as a function of time hence *actio* should be
taken over time.

$$a = msv$$
$$a/t = mvs/t$$
$$a/t = mv^2$$

What the concept of *actio* allows Leibniz to do is not only include time as a variable in his analysis of the measure of *vis*, it also allows him to rearticulate the generality of the concept of *vis* from another perspective. Until this point, Leibniz's
methodology for treating motion drew from a combination of Galileo's law of falling bodies and an extension of statics. In a mechanical system of constrained bodies, proportions between masses and motion were analogized to the static scenario
of the center of gravity. This methodological limitation means that the conservation
of mv^2 in unconstrained "natural" motion could be made to "fit" the statical scenario
through the use of the center of gravity or other devices (the mechanical raising of
a body before freefall). The turn to *actio* was a methodological change in this regard.

Take an inertial motion in an inertial system of a single body, with mass m, moving at velocity v. In such a system, the velocity of the inertial motion does not change. Hence the distance that the body traverses will be the product of its velocity and duration: $a/t = ms/t\cdot s/t = mv^2$. As duration increases, the distance traversed increases, and that proportion in time is simply velocity. The behavior of that body in time multiplied by its velocity at any time of measurement will yield mv^2 for the reasons described above. Now, we take the kind of scenario with which Catelan tried to criticize Leibniz's *Brevis Demonstratio*. In the upswing of a pendulum the speed of a body is exhausted in the transformation of speed into height, but this transformation of speed is proportional to the height (proportional to the achievement of formal effect).

$$a = ms \cdot v$$

$$a / \Delta v \propto \Delta ms$$

At maximum height, speed is nil, and the proportion is indefinite. But we see that just before it reaches that point, the exhausted speed of the upward swing is proportionally compensated by the effect that the motion has accomplished. The product of the proportionally increasing "formal effect" and the proportionally diminishing velocity gives us the two factors ($1/\Delta v$ and Δs) which, taken over equal time, will constitute the actual measure mv^2.[15]

The account of *actio* in the *Dynamica* does not appropriately account for the dynamics of the curvilineal motion of pendulum (see Chap. 3). We are required to extrapolate from the methodology. Nonetheless, we can see how this treatment of the transformation of velocity or speed in time begins to enter the territory of the later Lagrangian dynamics at least as a general methodological direction. In the first case, we have constant velocity and a constant ratio between distance traversed and duration. In the second case we have a diminishing speed and an increasing ratio between formal effect over the change of speed. Note that, although the formal effect in both the first and second case increases with duration (in time), their ratio to the transformation of the velocity of the body at the time of measurement is different. In time, as velocity diminishes, formal effect increases providing the proportion $\Delta a/v \propto \Delta s$.

In this analysis, we see what the introduction of the concept of action allowed Leibniz to do. The concept of causation is modified in the dynamics. This occurs in two steps. First, Leibniz generalizes the measure of mv^2 for force across locomotive phenomenon of different kinds. As we have seen, this allows Leibniz to make clear how violent and natural motions can be understood according to the same principle. The conceptual move necessary for this generality is to flatten the potential-actual model of causality. Without this conceptual move, inertial motion, collision, and mechanically constrained motion (like the pendulum) could not be understood under

---

[15] In Leibniz's later explication of the concept of *actio* in correspondence with Hermann, he is careful to make such a distinction between the two factors of evolving speeds and how *potentia* is constituted by *actio* over time (GM IV 388–389).

the same causal framework. Secondly, it follows from the above that the problem of the virtuality of "future effect" within the action of a body in motion is rendered actual. That is, although all future "effects" of motion remain in the future, this virtual future is actualized within a present action that is constituted by the actual velocity and achieved formal effect. This is nicely captured by the fact that $a/t = mv^2$. The distribution of *actio* in time is measured by *vis* qua mv^2. Indeed, as Leibniz will later plainly state in a 1699 letter to Bayle, "Action is nothing but the exercise of force and is represented by the product of force and time" (A II 3, 556). Hence, the invariance of action in time renders the exhaustion of *potentia* (of a mechanical system in time) only a derivative species of dynamical causality. *Vis* qua cause becomes a constant quantity, and its activity becomes a constant (invariant) actuality rather than the activity that brings a capacity or potential into actuality. It remains an open question whether this mathematical account of *vis* qua *actio* is a result of a conceptual movement away from the potential-actual model or the development of the mathematical account. The fact that the *Phoranomus*, which relied on the potential-actual model, was abandoned while the section on the (draft) mathematical formulation of the eventual *actio* was left unfinished suggests that this was what motivated Leibniz to move from the *Phoranomus* project to the *Dynamica* project.

With these two aspects of the concept of *actio*, generality and actuality, we have a new concept of causation. The aspect of generality allows us to understand the cause of motion as indifferent to a mechanical event (impact, pressure, mechanical constraint) or mere inertia (*quantum in se est*). Causation is thus understood as a formal principle governing motion as such. The aspect of actuality allows us to understand the cause of motion as constant or invariant. It is thus not a translation of potentiality to actuality but rather the activity of *vis* in time. Again, this makes causation a formal principle through the action of *vis* in time.

The metaphysics of formal causation will be addressed in a later chapter (Chap. 5). For the moment, I only want to emphasize the structural nature of this causation. There is no doubt that the measurement of *vis* qua mv^2 attributes a structural property to motion. This was the limit of Leibniz's methodology of measurement until the late 1680s. We examined this in earlier sections of this chapter. The measure mv^2 was gained through the proportion $h \propto v^2$, a measurement of *vis* through the proportion of work and maximum velocity. In this, *vis* was a property of the mechanical system as a whole. The extended static methodology employed by Leibniz allowed him to treat *vis* as an invariant for the proportion between work and motion. Though this Leibnizian concept of *vis* can only be understood as the property of the structure rather than of a body or a physical event, this is not yet structural causation. The measure of *vis*, mv^2, is a structural property, a property given through the invariant of a mechanical system. With *actio*, mv^2 remains a structural property but with the additional implication of the structural nature of the cause of motion. More than a quantity derived through measurement, motion is itself caused by *actio* in time. This action is of course distributed between bodies and across time. *Vis* is cause in a structural sense because it translates *actio* into formal effect and velocity.

The concept of structural causality is counterintuitive in a number of different ways. It replaces our intuitive sense of mechanical processes and interactions (exhaustion of potential energy, impact, etc.) with a two-tiered level of causation. There is a non-phenomenal reality of *vis* that is translated by *actio* in time as phenomenal effect. What is causal and what is caused exist on different levels of reality. Appealing to our analogy employed at the beginning of this chapter, we can say that *vis* is the Apollonian cone, and the effects of *vis* (in time) are like the many two-dimensional curves generated out of the structure. My aim here is not to defend the correctness of Leibniz's dynamics or its inherent theory of causation. It is rather to use the concept of structural causation to understand the dynamics project and to present this in a contextual and systematic way. This counter-intuitiveness can thus be mitigated by understanding Leibniz's arguments in the dynamics. In what follows, I will turn to Leibniz's last work within the dynamics project, the *Essay de dynamique* (circa 1700–1701), in order to take a final synthetic look at how this structural causation operates at the end of Leibniz's dynamics research.

2.4 *Actio, Vis*, and Causation in *Essay de dynamique*

The *Essay de dynamique* of (*circa*) 1700–1701, not to be confused with text of the same name (also in French) of 1692, was Leibniz's last attempt at a treatise-style presentation of the dynamics.[16] With the more comprehensive text of the *Dynamica* still waiting in the wings, Leibniz once again aims to argue from the perspective of quantity conservation, resuming his strategy of refuting Cartesian quantity of motion. Yet, Leibniz has gained needed profundity in his methodology, and his argument no longer resembles his earlier *Brevis demonstratio*. The concept of *actio* now plays a central role in the articulation of his theory and refutation.

The opening salvo of the *Essay de dynamique* once again restates the rejection of the Cartesian quantity of motion but quickly moves to distinguish this from another conservation concept, the quantity of progress (GM VI 216–217). The quantity of progress is nothing but the rectified Cartesian quantity of motion (with directionality). The distinction he makes is that, although the quantity of progress is indeed conserved, it is not conserved absolutely. Here, as we have mentioned above, Leibniz gives place to the conservation of the quantity of progress, the product of mass and velocity (mv), while limiting its scope. This quantity is not absolute because it only applies to constrained (and inertial) motion, whereas it does not apply to what he calls "violent" motion, i.e., freefall and work. The conservation of *vis* (mv^2) applies to both cases, but conservation of progress (mv) only applies to the limited case of inertial motion and rectilinear collision. The aim of the text is thus to show the difference between the two cases and the superiority and universality of the conservation of *vis*, mv^2.

---

[16] The *Essay de dynamique* in question is edited in Gerhardt's edition of the mathematical writings (GM VI 215–231). The 1692 *Essay de dynamique* is not edited in the *Gerhardt* or *Akademie Edition*, but we have a dutifully edited and annotated edition by Pierre Costabel (Costabel 1960).

Although this argument from generality has already been evident in his earlier work, the *Essay de dynamique* shows Leibniz's explicit commitment to this strategy and makes evident the evolution of methodology that Leibniz undertook in the preceding decade. The key methodological point is made very clearly. "[T]he one who confounds *vis* [*la force*] with the quantity of motion is abusing the doctrine of statics" (GM VI 218). The abuse in question is the transposing of the static doctrine of equilibrium between mass and distance to fulcrum onto the relation between mass and velocity with respect to the center of gravity. What is being abused is not the symmetry (with respect to the center of gravity) provided by statics, which holds, but rather the identification of this symmetry to the quantity of motion (or progress).

Leibniz here relies on an idiosyncratic distinction between "violent" and inertial motion to distinguish different cases of symmetry. Leibniz's counterpoint to "violent" motion is "harmless" motion, as we find it in *Specimen dynamicum* (GM VI 243; AG 127). The term "violent" (counterposed to "natural") motion comes from the Aristotelian tradition, where a distinction is made between motion resulting from the un-interfered motion of a body (like the earth) finding its place in the center of the earth and the mechanical motion of a body (like the earth) thrown by a machine. Despite Leibniz's sympathies with some aspects of Aristotelian physics, his use of "violent" and "harmless" have little to do with the earlier context. Despite the idiosyncratic terminology, it is clear that the distinction between "violent" and "harmless" correspond to non-inertial and inertial motion. It is in non-inertial motion that the conservation of *vis* (or energy) is conspicuous.[17]

With this distinction Leibniz makes the case for the generality of the conservation of *vis* through the concept and conservation of *actio*. After our examination above, the argument should appear rather simple. Recall that *actio* is divided into two factors. The first is formal effect (in time) and the second the velocity at the time of measuring formal effect. In inertial motion, formal effect over time is simply mass and velocity (ms/t), and the product of that factor with velocity is simply mv^2. Things are different in non-inertial motion. In "violent" motion, like the upswing of a pendulum, velocity decreases as formal effect in time increases. Symmetrically, in cases of free fall, formal effect over time decreases as velocity increases in time.[18] Here Leibniz uses the terminology of "violent effect." In the earlier model where *potentia* is exhausted in motion, the accomplishment of violent effect produces the measure of conserved *vis* is made through the equipollence of full effect and entire

---

[17] Speaking of inertial and non-inertial motion should not be confounded with inertial and non-inertial systems or reference frame. The former relates to unaccelerated and accelerated motion, but both can be registered inside an inertial reference frame (or system). Non-inertial systems refer to systems where the frame itself is accelerated with respect to some inertial frame. My thanks to the anonymous reviewer for flagging this possible ambiguity.

[18] This statement is misleading if we fixate on the fact that formal effect increases in absolute terms in time. Since a body accelerates in freefall (regardless of the rate of this acceleration), the path covered by the falling body over time decreases in proportion to the speed of the accelerating body at the time of measurement. In other words, each new state of acceleration is at a faster rate than the average speed calculated in the fall before the time of measurement.

cause. This relation between violent effect and the diminishing velocity certainly reproduces mv^2 but Leibniz's aim here goes further.

Now, before examining Leibniz's case study in the *Essay de dynamique*, it is important to underline that Leibniz makes the argument that "absolute force" should be measured by "violent" rather than "harmless" effect. The reason here has been examined above. Since the process of a "violent" effect "consumes" or exhausts the power of a moving body, it is in such cases, as opposed to inertial motion, that the measure of *vis* becomes apparent. If we hold that the conservation of *vis* qua mv^2 is a structural property, then we require the occasion of "violent effect" to make clear the difference between the conservation of the quantity of progress and the quantity of *vis*. But the method through which this difference is made in the *Essay de dynamique* illustrates a more profound consideration provided by the concept of *actio*. Leibniz states that the quantity of *actio* is invariant during the intervals of time separating measurements (GM VI 220). We know that *vis* is conserved in any system, and this is certainly obvious in inertial systems. However, the conservation of *vis* within the earlier conception can only be rendered without time. The proportion $h \propto v^2$ is only evident in statical terms, taking maximal height and maximal velocity. This is part and parcel of a methodology of the equipollence of cause and effect, a cause (*vis*) and its future effect (a body raised to a height). The concept of action provides the term through which the conservation of force can be expressed as an invariant across time and across different sorts ("harmless" and "violent") physical systems.

Leibniz's case study and demonstration in the *Essay de dynamique* provides a rare occasion for directly applying the concept of *actio* in the examination of a physical system. The aim, as we have mentioned, is to show that quantity of progress is not conserved universally while quantity of *vis* is. Where this differs from other texts is that Leibniz uses the quantity of *actio* to show the conservation of the quantity of *vis*. Now, despite his previous remarks concerning the distinction between violent and "harmless" action, Leibniz refers his readers to other texts for the explication of the conspicuous relationship between absolute *vis* and violent effect.[19] Here instead, Leibniz focuses his attention on inertial motion and collisions. Our analysis above already provides the context for the strategy Leibniz pursued here. Since the translation between violent effect (the exhaustion or "consumption" of intensive *potentia* in motion) and *vis* is the obvious case, Leibniz treats the more subtle case of inertial motion (and collision) in order to show the explanatory power of the concept of *actio* for the measure of force.

Leibniz's case study in the *Essay de dynamique* is complex, and I have simplified it here with the use of figures and charts. There are two stages, divided between two time intervals (from t_1 to t_2 and from t_3 to t_4) (GM VI 224–225). First, there is an initial system of three bodies, a, b, and c, in oblique linear collision. From time t_1 to t_2, the body a moves with a positive velocity to strike two stationary bodies b and c each at half-right angles (45°). The two bodies b and c are thus rebounded along axes L and M at a right angle. At the second stage, from time t_3 to t_4, there are other bodies along the axes L and M, d (on axis L), and e (on axis M). Since b and c are rebounded and travel along the axes L and M, they encounter bodies d and e respec-

---

[19] See GM VI 217–218 and GM VI 117–119.

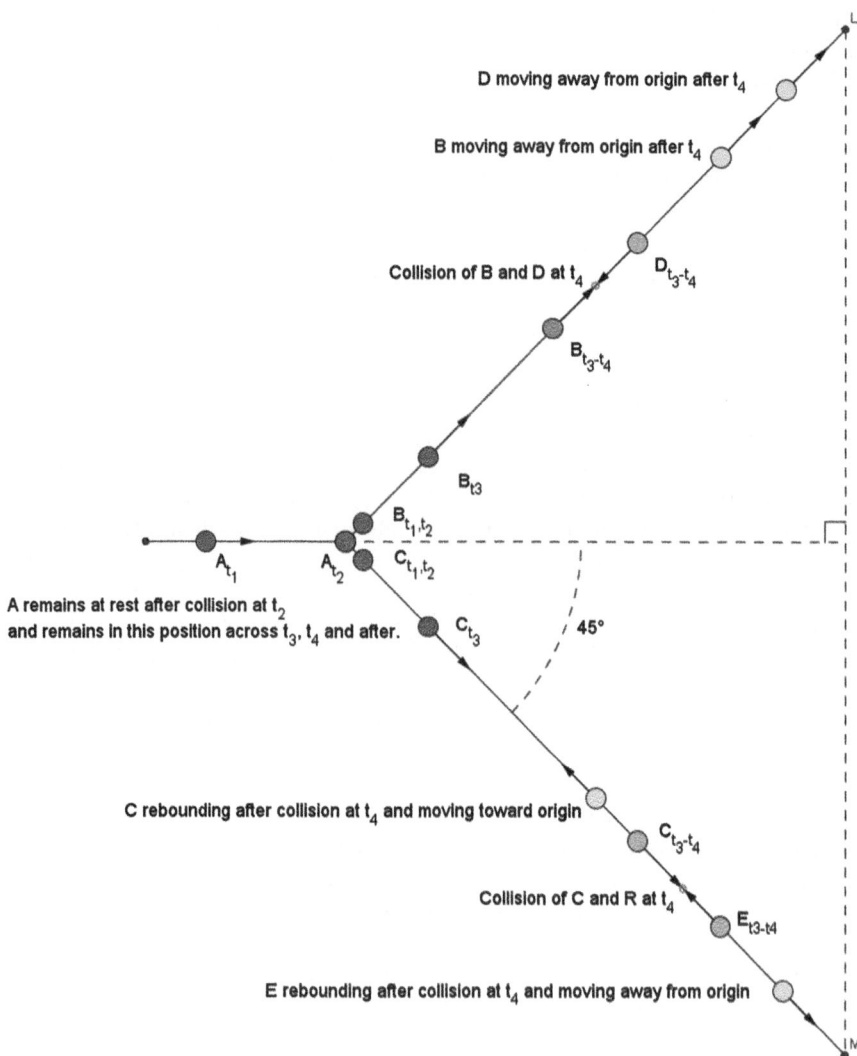

Fig. 2.5 Motions and collisions of bodies A, B, C, D, E at times t_1, t_2, t_3, t_4 in *Essay de dynamique*

tively and engender rectilinear collisions and change of velocity. The following chart shows the two stages of the case study (Fig. 2.5).

A few things should be highlighted here. First, the collision of the body a (moving at a velocity of $\sqrt{2}$) on the two bodies b and c at half-right angles allows the bodies b and c to be repelled each with a velocity of 1. The most intuitive way of understanding this is through the conservation of momentum. The problem of oblique collisions has been treated in Leibniz's correspondence with Bernoulli and De Volder on 27 December 1698 (GP II 159–169; LdV 39–43). The issue in the correspondence concerned the question of the velocity of the two bodies in oblique, half-right angle collisions. In the case of the correspondence, as well as in the *Essay de dynamique*, Leibniz makes use of the trigonometrical relationship to divide the impact of a on b and c into the resulting velocities. Although Leibniz does not mention it, this geometrical relation in impact preserves both the symmetry of momentum (as well as *vis*). This conservation of momentum can be quickly illustrated from the analysis of the impact from vertical and horizontal components:

$$m_a v_a = m_b v_b \cos\theta_1 + m_c v_c \cos\theta_2 + m_b v_b \sin\theta_1 - m_c v_c \sin\theta_2$$
$$m_a v_a = \left(1 \cdot v_b \cdot \sqrt{2}/2 + 1 \cdot v_c \cdot \sqrt{2}/2\right) + \left(1 \cdot v_b \cdot \sqrt{2}/2 - 1 \cdot v_c \cdot \sqrt{2}/2\right)$$
$$m_a v_a = v_b \sqrt{2}/2 + v_c \cdot \sqrt{2}/2$$

Given the symmetry of the collision, $v_b = v_c$, hence:

$$m_a v_a = 1 \cdot \sqrt{2}/2 + 1 \cdot \sqrt{2}/2 = \sqrt{2}$$

Hence, momentum is conserved.

Here Leibniz exploits the same trigonometric relation to deliver the measure of *vis*. Since the *vis* of body a is 2 (or $m_a v_a^2 = 1 \cdot \sqrt{2}^2$), the power it imparts on each body (a and b), upon impact, is 1 respectively.

The second thing to highlight is that the body c rebounds after striking body e at time t_3, but b and d do not rebound after colliding. Although Leibniz addresses elastic and inelastic collisions later in the text, he does not specifically address this issue. However, judging from the fact that *vis* is conserved throughout, all the collisions are elastic. It appears that Leibniz wanted to incorporate different kinds of motion within his example (Fig. 2.6).

This chart reproduces Leibniz's account of the case study in order to demonstrate that the quantity of motion is not conserved while action in time, and hence *vis*, is. Of course, Leibniz exploits the oblique collision at the first stage of the case study ($t_1 - t_2$) in order to disqualify the quantity of motion as the conserved quantity. What follows in the second stage of the evolution of the system ($t_3 - t_4$) are all rectilinear collisions. Now, despite Leibniz's explicit statement, going into the case study, that quantity of progress is conserved in some cases (in linear collisions), it must be emphasized that momentum, if properly measured through its angular components, can also be shown to be conserved. This shows a limitation in Leibniz's methodology, of course, namely the lack of sophistication in handling the conservation of momentum through the elements he certainly had at his disposal: directionality and trigonometry.

Indeed, in retrospect, one could easily write off the *Essay de dynamique* for the fact that the central argument is established on the basis of a possibly deliberate misunderstanding. However, this limitation reveals something central to Leibniz's general methodology. The fact is that Leibniz's concept of *vis* does not cover the relationship between collision and the transformation it engenders. Rather, its limitation is due to the narrow focus placed on invariance across collisions and motions.

		A	B	C	D	E
	mass	1	1	1	2	½
t_1	velocity	0	0	0	0	0
	formal effect=mass·distance	0	0	0	0	0
	actio=formal effect·|velocity|	0	0	0	0	0
t_2	velocity	$\sqrt{2}$	0	0	-½	-2/3
	formal effect=mass·distance	$\sqrt{2}$	0	0	1	2/6
	actio=formal effect·|velocity|	2	0	0	½	2/9
t_3	velocity	0	1	1	- ½	-2/3
	formal effect=mass·distance	0	1	1	1	2/6
	actio=formal effect·|velocity|	0	1	1	½	2/9
t_4	velocity	0	1/3	-1/9	5/6	14/9
	formal effect=mass·distance	0	1/3	1/9	5/3	7/9
	actio=formal effect·|velocity|	0	1/9	1/81	25/18	98/81

Fig. 2.6 Table of variables for bodies A, B, C, D, E at times t_1, t_2, t_3, t_4 in *Essay de dynamique*

Sum of various quantities for bodies A, B, C, D and E.			
	formal effect	actio	mv^2
t_1	0	0	0
t_2	4/3·√2	49/18	49/18
t_3	10/3	49/18	49/18
t_4	26/9	49/18	49/18

Fig. 2.7 Sums of various quantities for bodies A, B, C, D, E at times t_1, t_2, t_3, t_4 in *Essay de dynamique*

The analysis of this case study through the concept of *actio* renders this apparent precisely through its severe gaps (Fig. 2.7).

This chart, based on the data of the previous one, allows us to focus more clearly on this invariance (*actio* or *vis*) at work in this example. What it illustrates is clearly the discrepancy between formal effect (ms) and action as well as the invariance of *actio* (and *vis*). Referring to the previous chart, we also see that this tabulation includes, in the first stage of the test case ($t_1 - t_2$), the motion of bodies d and e which have not come into any mechanical relationship at all with the initial system of bodies a, b, and c. The "participation" of these bodies d and e into the more restrained system of the three bodies in collision constitutes the systematic quantity of *vis*. These bodies stand in for the many other bodies—the entire universe, in fact—that could be added to the physical system under consideration. From this, we see that Leibniz is concerned about a concept of *vis* that is systematic in nature.

In the *Essay de dynamique* we find that Leibniz uses the concept of *actio* to provide us with a theory of *vis* that is causal in a structural way. We have now moved far away from a methodology that relied on the concept of power based either on exhaustion or impact. The methodology of the measurement of *vis* is now based on the invariance of *actio* in a physical system with arbitrary number of bodies and the selection of an arbitrary (inertial) reference frame. More than this, the invariance of *actio* across time, such as we have analyzed above, also guarantees that this system can be generalized not only in space but also in time.

The aim of this examination is not only to trace the many ways that Leibniz matured in his methodology for the measure of *vis*. The true aim is to show how the concept of *vis* matured through these methodological changes. What occurs through the development of the dynamics project is the transformation of a methodology based almost exclusively on *potentia*, the exhaustion of an intensive quantity into an extensive quantity, to an invariant capable of subsuming the concept of *potentia*, inertial motion, and even larger physical systems with no mechanical interactions at all. Of course, as in many cases with Leibniz, this maturation was not linear. The elements of his final expression in the *Essay de dynamique* were already available to him much earlier. Certainly, it is easier to say that the quantity of *vis* is what is conserved by God in the universe and much harder to show how this conservation operates (in space and time).

2.5 Concluding Remarks

In this chapter, we have initially taken up the causality of *vis* as understood through *potentia*, as a structural property. That is to say, the causal nature of *vis* is understood through the quantity mv^2 which is a predicate of a physical system. This aspect of the dynamics remains a key aspect of the concept of *vis* until the end of the project. Of course it is hard to see how a property like a conserved quantity can be causal. However, Leibniz's dynamics presents a theory of causation that puts *vis* in a role that is much more important than as a predicate of a structure. Of course, the very idea of the dynamics is the examination of motion from the perspective of its cause. The introduction of the concept of *actio* allows *vis* to be dissected into different temporal parts. It is hence this concept *actio* that allows us to grasp how *vis* is causal.

Given our examination above, *vis* is causal through its invariance and *actio* is what allows this invariance to be translated into motion in space and time. This causation is thus structural because it does not depend on impact, the exhaustion of an intensive quantity or any other mechanical process. Rather, this form of causation subsumes these different mechanical processes. In short *actio* governs the translation of cause (*vis*) into effect (motion) through the distribution $a/t = msv/t$. The conservation quantity mv^2 is not causation itself, but it demonstrates the governing framework of this causation. As such, dynamical causation can be understood as structural causation. Rather than a relation between mechanical powers, dynamical causation is the *actio* of a physical system at every moment of its evolution and also integrally so. As such, dynamical causation is the relation between *vis* (qua cause) and phenomena.

The interpretation of dynamical causation as structural causation depends on a distinction between levels of reality since *vis* is non-extensional and non- or infra-phenomenal but nonetheless causes extended locomotive phenomena. In a certain sense, one might say that phenomena are "less real" than *vis*. Of course, the cause-effect relation means that phenomena are derivative but no less real. Since *vires* are real, what they cause, phenomena, are also real. If *vires* are "real" then their effects are "real" as well. There is no reason to treat the distinction of different levels of reality other than for the sake of its underlying methodology.

As a conclusion to this chapter, I wish to reiterate the central development of the above argument in order to solidify the lens through which Leibniz's dynamics will be understood in the rest of this book. The interpretational idea here is that the development of Leibniz's methodology in treating the conserved quantity of *vis* mv^2 leads to a theory of structural causation. The second section of this chapter outlined Leibniz's sources and the reasons for the establishment of the quantity mv^2 as the foundation of his research in mechanics. Initially posed against the Cartesians, mv^2 was argued to possess a greater generality than the rectified Cartesian quantity of motion. Here, even if Leibniz could accept the conservation of the rectified Cartesian quantity of motion in some cases like rectilinear collision, it remained limited. The quantity mv^2 was thus posed as the superior candidate for the conserved quantity in nature. Its measurement followed the concept of *vis* as a structural *property*, a property of a physical system taken integrally. In the third section, we examined the introduction of *actio* into the relation between *vis* and its extensional effects (motion). Here, the structural property of *vis* is dissected into the two factors, distance traversed in time and speed. The actuality of *actio* accounts for both of these factors and provides the proportion between the temporal translation between an extended motion and an intensity (speed) that is yet to be extended. Dynamical causation thus becomes, through *actio*, the translation of an atemporal and non-spatial entity in space and time. The fourth section of this chapter attempted to make this counter-intuitive idea concrete in the case study provided in the *Essay de dynamique*. In brief, dynamical causation is structural because it causes locomotive phenomena through the organization (or structuring) of this phenomenon.

The aim of this chapter was to reveal the character of dynamical causation through the development of methodology in Leibniz's dynamics. The central emphasis was the transition from a methodology based on *potentia* and a methodology based on *actio*. The methodological difference, as we examined above, is clear. That is, the generality of *actio* for the reckoning of the measure of *vis* transcends and subsumes mechanical interaction. The conceptual difference here however is that causation cannot be merely understood as mechanical relation but rather as a structural action and cause.

Chapter 3
The Equivalence of Hypotheses and Dynamical Causation

Abstract The third chapter of this book begins a three-part presentation of the central architectonic components of the dynamics (continued in the following chapters). In this chapter, we examine the so-called equivalence of hypothesis, inherited from Kepler, Galileo and Huygens, and see how it shaped the concept of force and the method of its measurement. The main aim is to show the theoretical independence of the equivalence of hypotheses and the theory of Leibnzian *vis* in order to understand how the former principle provides the limits of physical phenomena. It is within these limits that Leibniz *vis* can be understood as cause.

3.1 Introduction

The previous chapter provided a presentation of the dynamics through a synoptic view of the theory of dynamical causation as structural causation. In the next three chapters, I will examine three important architectonic components of the dynamics and demonstrate their contribution to the theory of structural causation. In this chapter, I will examine the equivalence of hypotheses and its relation to the concepts of *vis*, *actio*, and motion. I will then examine the principle of continuity as it relates to the dynamics and then the principle of the equipollence of cause and effect in Chaps. 4 and 5, respectively.

The equivalence of hypotheses was a principle that Leibniz inherited from the controversies around Copernicanism. For centuries before Copernicus, the flaws of Ptolemaic astronomy were compensated with elaborate supplemental theories, especially with regard to the movement of moons. As the story goes, astronomical tables built from centuries of observation provided the data to fashion principles that make observed celestial movements consistent with these elaborations of the Ptolemaic theory. The theoretical superiority of Copernicus' heliocentrism was not immediately evident because the explanatory power of the new (or renewed) theory still lagged behind the centuries of accumulated data and methods available to the traditional Ptolemaic account. This gap between theory and observation was addressed by Kepler who, in the *Astronomia Nova*, recognized the geometrical equivalence of the Ptolemaic, Tychonic, and Copernican models. An incomplete or erroneous causal theory could nevertheless produce adequate geometrical

© Springer International Publishing AG 2017   41
T. Tho, *Vis Vim Vi: Declinations of Force in Leibniz's Dynamics*, Studies in
History and Philosophy of Science 46, DOI 10.1007/978-3-319-59055-4_3

explanations (Kepler III 1990, 71–77, 87–106; 1992, 130–139, 155–180).[1] It is hence important to pay attention to causes, or to attend to a theory of causes, rather than be satisfied with geometrical adequation (Kepler 1984).

Indeed, in many places Leibniz invokes this Keplerian view of the inadequacy of geometrical hypotheses to give real determination to celestial and mundane loco-motion. But despite this and his expressions of admiration for the astronomer, Leibniz extended Kepler's notion of the geometrical undecidability of physical cause to the very conception of motion itself. Leibniz takes this to the extreme when, even though he was clearly a partisan of heliocentrism, he denies the abso-luteness of either geocentrism or heliocentrism, instead providing a pragmatic crite-rion for the "truth" of these cosmological hypotheses. As Leibniz puts it in a 1689 text erroneously entitled *Phoranomus seu de potentia et legibus naturae* by L. Couturat and retitled *On Copernicanism and the relativity of motion* by R. Ariew and Daniel Garber,

> [I]n explaining the theory of planets, the Copernican hypothesis wonderfully illustrates the soul, and beautifully displays the harmony of things at the same time as it shows the wis-dom of the creator, and since other hypotheses are burdened with innumerable complexities and confuse everything in astonishing ways, we must say that, just as the Ptolemaic account is the truest one in spherical astronomy, on the other hand the Copernican account is the truest theory, that is, the most intelligible theory and the only one capable of an explanation sufficient for a person of sound reason. [...] [T]he truth of a hypothesis should be taken to be nothing but its greater intelligibility ... henceforth there would be no more distinction between those who prefer the Copernican system as the hypothesis more in agreement with the intellect, and those who defend it as the truth. For the nature of the matter is that the two claims are identical; nor should one look for a greater or a different truth here. (C 590–593; AG 90–94)

This is indeed yet another occasion for witnessing Leibniz's ecumenical disposition at work, but we shall put this question of philosophical "style" aside. What is key here is to recognize the degree to which Leibniz instrumentalizes and relativizes the cosmological theories on the basis of the equivalence of hypotheses.

Leibniz's generalization of the equivalence of hypotheses, a central tool for his examination not only of celestial motion but motion as such is the formalization of the well-known "Leibnizian relativism of motion" that had been the central point of a great number of commentaries on Leibniz's physics. What is this relativism? I agree here with the established consensus of the Leibniz community in reducing this to a theory of motion that is grounded on relative velocities with respect to an arbitrary (inertial) reference frame.[2] This is in fact nothing more and nothing less than what the principle of the equivalence of hypotheses states and conforms to so-called Galilean "relativity" or invariance. Although the terminology of "equiva-lence" is inherited from Kepler, for Leibniz, this principle states that any set of hypotheses concerning the velocity of motion for a physical system of bodies is equivalent as long as their relative velocities are respected and the frame is inertial (non-accelerated). For the sake of establishing a definition of the equivalence of

---

[1] See Voelkel 2001.

[2] See, for example, Arthur 1994.

hypotheses, let us take one of the main programmatic statements of the principle from the posthumously published second part of the *Specimen Dynamicum*:

> [W]e must hold that however many bodies might be in motion, one cannot infer from the phenomena which of them really has absolute and determinate motion or rest. Rather one can attribute rest to any one of them one may choose, and yet the same phenomena will result [...] [T]he equivalence of the hypothesis is not changed even by the collision of bodies with one another, and thus, that the laws of motion must be fixed in such a way that the relative nature of motion is preserved, so that one cannot tell, on the basis of phenomena resulting from collision, where there had been rest or determinate motion in an absolute sense before the collision. (GM VI 247; AG 131)

We shall examine the principle in more detail below, but we should briefly note a few things from this brief presentation of the principle of the equivalence of hypotheses. First, we should underline that it satisfies the criterion of the relational status of motion according to a chosen (inertial) reference frame. Take the hypothesis that a body a moving towards body b, at rest, at velocity v is equivalent to the hypothesis of the body b moving at velocity -v towards body a, at rest. All intermediate relative states of motion and rest between body a and b express the same motion whether a travels at 1/2v and b at −3/2v or a at 5/4v and b at 1/4v, given the shift in (inertial) reference. Hence, where u is the arbitrary non-accelerated rectilinear motion of the reference frame:

$$v'_a - u = -\left(v_a - u\right)$$

The mathematical continuity implied by this principle will be examined in the next chapter, but here we wish only to point to the concrete meaning of the relationism provided by this principle. Any variable of motion in a physical system is assigned only with respect to its relation to other variables.

The continuous relation of the transformation of variables of motion between individual bodies within any physical system leads us to the second key aspect of the principle of the equivalence of hypotheses. That is, the equivalence of hypotheses assures the identity of a physical system. In its original Keplerian context the principle had a negative valence. It marked an epistemic limit for the use of geometry to understand astronomical hypotheses. A major strategy of the *Astronomia nova* exploited Kepler's demonstration that "these three forms [Ptolemy, Copernicus, Brahe] are absolutely, perfectly, geometrically equivalent," for the sake of identifying "physical causes" later in his treatise (Kepler III 1990, 89, 288–290; 1992, 157, 455–458).[3] The idea here is that all these "phenomena-saving" geometrical descriptions were not able, in themselves, to determine their unique veracity but, insofar as they are each coherent, they are all able to establish their own plausibility. A solid astronomical method therefore required more fundamental causal "physical hypotheses" to supplement geometry. Leibniz's use of the principle, based on a general theory of relational motion, transforms this epistemic limit of geometry into a delimiting feature of physical phenomenon. As we cited earlier, Leibniz remarks that the continuous transforma-

---

[3] On Kepler's method see Itokazu 2009.

tion of the relative velocity of motions in a physical system does not change the identity of the phenomenon. More than this, Leibniz adds that this is not changed in the evolution of the physical system, even in the case of collision. The identity of a phenomenon is thus an invariance that is constituted in and through its variation.

These two aspects of the equivalence of hypotheses highlight another aspect of the theory of structural causation that was not addressed in the previous chapter. Now, in the previous chapter, we argued that the causal nature of *vis* is to be understood through the translation of a cause to a spatial-temporal system of phenomena (a physical system). Such a causation is structural because it is the action of force expressed through its invariance while deployed in locomotion. Leibniz's use of the principle of the equivalence of hypotheses then deepens our understanding of this translation. This principle allows for the same cause to be expressed by an infinite set of internally related motions.[4] This is simply the group of velocities that occurs through the arbitrary shift of (inertial) reference. For any given cause (qua *vis*), a continuous group of velocity variables, given that they respect relative motion, satisfies the effect. Hence the relationship between cause and effect is between *vis* and a group of internally related motions. This positive use of the principle of the equivalence of hypotheses therefore underpins a crucial aspect of the theory of structural causation.

The aim of this chapter is to deepen our understanding of the dynamics through examining how structural causation is reflected in the principle of the equivalence of hypotheses. The main challenge here is to demonstrate the independence of the equivalence of hypotheses and the core theory of *vires*. The idea here is that, while the central theory of *vires* provides an account of the translation of cause to effect, the equivalence of hypotheses provides the delimitation for the form of this effect, the limits of phenomena. This is a crucial aspect of the theory of structural causation because it makes it possible for us to account for the phenomenon caused by structural causation (See Chap. 2).

Given this aim, this chapter will examine next, in the second section, the conditions of this generalized principle of the equivalence of hypotheses. As we touched on briefly above, Leibniz's use of this principle is far from its original Keplerian context. Instead of being merely a principle of the limits of epistemic access, it becomes a positive deployment of Galilean and Huygensian "relativity of motion" as a methodology for the interpretation of variables in measurement. In the third section, through an examination of motion relationism's greatest challenge, rotational motion, I demonstrate Leibniz's attempt to sweep away the problem through an appeal to the limits of physical phenomenon. Here we will establish how Leibniz treated the equivalence of hypotheses not only as a methodology for measurement but also a theory of the form of locomotive phenomena. The key here is to argue for the independence of the equivalence of hypotheses from the theory of *vires*. If a physical phenomenon is constituted by an invariance governing variation, the principle of the equivalence of hypothesis is not itself sufficient to provide this principle of invariance. A theory of *vis* is needed beyond what the equivalence of

---

[4]The set of mutually related motions is infinite because we assume that the equivalence of hypotheses allows for continuous transformations.

hypotheses could provide. In the fourth section, I will clarify how the form of loco-motive phenomena established through the equivalence of hypotheses is different from the eventual concept of inertia. This will allow us to determine the role played by the equivalence of hypothesis. Finally, I will provide an account of the synthesis of a theory of *vires* and a theory of phenomenon in the constitution of the concept of causation in the dynamics. The principle of the equivalence of hypotheses thus provides another way to understand how *vis* is translated into motion. It provides the principle of variation. This chapter thus provides an examination of the principle of the equivalence of hypotheses in order to complete the general exposition of the concept of structural causation.

3.2   Absolute and Relational Motion

Leibniz's relationism about motion is perhaps the most discussed and disputed aspect of his physical theory. However, this relationism refers to two separate doctrines. The first, which we will not discuss in this chapter, is the ontological relationism of motion. This refers to the ontological dependence of space, and hence, extended motion, on the *existence* of bodies. If bodies were not created, there would be no motion and hence no space. This is a complicated issue and will be addressed in Chap. 6. Here I only emphasize that this is a problem of ontological dependence of motion on bodies and does not (directly) interfere with the structures of space or motion themselves. The second meaning of Leibniz's relationism is precisely the equivalence of hypoth-eses. The principle can be simply understood as invariance under the shift of (inertial) reference frames. As we have explained above, this can be formalized as:

$$v'_a - u = -\left(v_a - u\right)$$

In Leibniz's many explanations of this aspect of his relationism of motion, he appeals to his mentor Huygens' work on rectilinear collisions. To understand the extent and limits of his appeal, it is necessary to examine how Huygens' theory of collisions provided Leibniz with the principle of the equivalence of hypotheses.

As we noted in the last chapter, Huygens had already formulated the principles of collision and the mv^2 conservation law as early as 1656. The full treatise of *De motu corporum ex percussione* was made available only in 1703 in the *Opuscula postuma*, after Huygens' death in 1695. Whether Leibniz had access to the full trea-tise is unknown, but he did have a working knowledge of Huygens' methodology. We know this because, already in the 1678 *De corporum concursu*, Leibniz had a position similar to Huygens' argument of the treatise already in the 1678. Leibniz also makes reference to direct conversations between him and Huygens in Paris about the symmetry of collision in later correspondences (A III 6, 131; A III 6, 162). Nonetheless, the aspect of Huygens' theory that Huygens himself made public in the two publications in 1669 (in the *Philosophical Transactions* and the *Journal de*

Fig. 3.1 Illustration of Huygens' "boaters" experiment for relative motion (Image taken from Huygens 1929, 29)

sçavans), namely the theory of motion and collision in an arbitrary reference (inertial) frame, that became the first part of the *De motu corporum ex percussione* was certainly known to Leibniz.[5]

Let us then set the stage for an examination of Leibniz's use of the Huygensian theory. The theory of the motion in an arbitrary reference frame was presented by Huygens through a thought experiment of a rectilinear collision occurring on a boat moving in a stream (Fig. 3.1).

The motion will appear different from the experimenter on the boat than on the shore. Indeed, if the boat itself is moving at a velocity -u, the experimenter on the boat sees body a moving at v and body b moving at -u, the experimenter on shore will see body a at rest and body b moving at −2v. This gives us the principle of Galilean relativity that we have already discussed earlier:

$$\text{Since } v'_a - u = -\left(v_a - u\right) \text{if } u = 1/2\left(v_a + v'_a\right)$$

Huygens used this principle of Galilean relativity to directly demonstrate velocity reversal in elastic collision. That is, using arbitrary changes of the velocity of the boat, the experience of the experimenter on the shore can produce a range of velocities for bodies a and b provided that their relative velocities are respected. Any difference between initial velocities between two bodies can be transformed into a symmetric collision with the appropriate motion of the reference frame (the boat). Now, given the case of elastic collision that is at work, any elastic collision between two bodies (of equal mass) can be reduced to a symmetric collision. Hence, this use of arbitrary reference frames deploys Galilean relativity in order to *demonstrate* the symmetry of collision:

---

[5]We should note that the use of "inertial frame" to describe Huygens' argument is technically anachronistic. The concept of inertia was insufficiently developed in Huygens' context. We also note that Leibniz was so inspired by reading Huygens' argument in 1669 that he copied the argument (rather than the arguments of Wallis and Wren) and then immediately composed a response to the article, noting his agreements and severe disagreements. Before his *Theoria motus abstracti* and *Theoria motus concreti* of 1671, Leibniz attempted to write a treatise on mechanics (in 1669) based on his response to Huygens' arguments. See A VI 2, 157–159, 161. His views were then discussed with in correspondence with Oldenburg in 28 September 1670. See A II 1, 101–102.

$$u = 1/2\left(v_a + v'_a\right)$$
$$v'_a - u = -\left(v_a - u\right)$$
$$v'_b - u = -\left(v_b - u\right)$$
$$\text{Hence, } v'_b - v'_b = -\left(v_a - v_b\right)$$

Employing this symmetry rendered by a clear application of Galilean relativity, Huygens moves to argue for his conservation principle mv^2 or $h\alpha v^2$.

We have seen in the previous chapter that the conservation of mv^2 holds for inertial and non-inertial motion, and that the measurement of this quantity is possible for each of these different cases provided the appropriate rule. However, this rule, the conservation principle, cannot be deduced from such an inertial framework alone. What is required is a version of Torricelli's principle, the principle that no physical system can raise its center of gravity. Huygens' inserts Torricelli's principle in his demonstration through the following experimental framework.

Starting with proposition 8 of *De motu corporum ex percussione*, Huygens argues that two bodies, A and B, whose masses are inversely related to their heights, dropped in free fall, and deflected on two 45 degree ramps, will collide with each other horizontally and deflect each other, climb back up the ramps and attain different heights. In this transformation, the center of mass of each body is maintained throughout and measured across three crucial moments: the initial drop, where height is measured; the moment of collision, where the position of collision is measured; and the final maximum height, where final height is measured again. Huygens' argument is that, if we assume, against hypothesis, that the greater mass of A will result in a lesser speed in A after collision, then the final heights of bodies A and B will produce a center of mass (O) higher than the initial center of mass (N). This is not the case. When A and B rebound and attain their respective heights, the relative velocity exchange will result in the same center of mass understood geometrically. Recall that the center of mass is determined by the statical relation between height, mass and distance to the center of mass frame. Now, Huygens does not provide this argument from empirical grounds, but rather *a priori* from what he had already established earlier in the treatise.[6] The proposition is to be read as an extension of his demonstration of center of mass in linear elastic collision earlier in the treatise. In strict terms the proposition is one that aims at showing why the center of gravity after collision is not raised to point O rather than a demonstration that the center of gravity is not raised to O, contrary to what the proposition explicitly states. I shall leave the discussion of the rigor of this methodology aside for the sake of emphasizing the content of Leibnizian inheritance of the methodology. Here the invariance established by center of mass provides the means by which we determine the motion of a system of colliding bodies in an arbitrary (inertial) reference frame (Fig. 3.2). For any two bodies A and B in collision, the quantity conserved (proportional to mv^2) before and after collision (marked by the prime) is:

---

[6] Earlier aspects of the treatise are included in summary in the derivation immediately below.

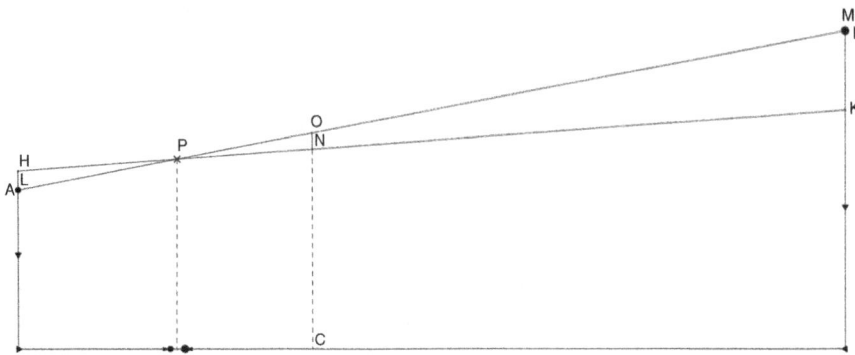

Fig. 3.2 Huygens' experiment for conservation under collision using center of gravity

$$m_A v_A^2 + m_B v_B^2 = m_A v_A'^2 + m_B v_B'^2$$

This follows first from Leibniz's laws of collision that conform to the Huygensian reversibility of relative velocity. As we have argued above, supposing u as the arbitrary motion of the (inertial) reference frame:

$$v_b' - v_b' = -(v_a - v_b)$$

As such, assuming a generic version of Galileo's law of falling bodies, where the duration of fall (proportional to height) of a body is proportional to the square of its velocity, it follows from the center of gravity argument and the above (algebraic) symmetry that:

$$m_A v_A^2 + m_B v_B^2 = m_A v_A'^2 + m_B v_B'^2$$

As we have already mentioned, this demonstration also allows us to infer the conservation of classical momentum:

$$m_A (v_A - v_A')(v_A + v_A') = -m_B (v_B - v_B')(v_B + v_B')$$

Now if we read Huygens' treatise as being divided between two parts, the first part, up to proposition 8, making use of the boat experiment to simulate arbitrary reference (inertial) frames, and the second part, after proposition 8, using the center of gravity to demonstrate conservation, we can neatly divide the two main results of Huygens' treatise. The first result is a theory of inertial motion (or proto-inertial motion) and the second result is a theory of the quantity conserved in collision. As we argued above, inertial motion is not itself sufficient to derive the conservation of mv². This is the crucial point for helping us understand why the equivalence of hypotheses must be independent from the theory of *vires*.

From the perspective of invariance and variation, the equivalence of hypotheses itself only provides us with the conservation of momentum as invariance. Of course, neither Huygens nor Leibniz emphasized this point, but Huygens' method (inherited by Leibniz) implies that any rectilinear elastic collision, with appropriate adjustments for differences in mass, can be reduced to a symmetric collision. Momentum conservation then follows rather directly. A demonstration for the invariance of mv^2 does not immediately follow from this, but we have seen how such an initial demonstration allows us to prove that the motion that converts vertical to horizontal motion allows us to arrive at the conservation of mv^2.

The examination here should make two things clear. First, with the separation between the equivalence of hypotheses and the theory of *vis* qua mv^2, we have two principles. The principle of the equivalence of hypotheses provides an account of relative velocities with respect to arbitrary (inertial) reference frames. The theory of *vis* provides a measure for a conserved quantity. The first is responsible for variation, the second is responsible for invariance. In this, the second principle, the theory of *vis*, provides an aspect to the dynamics which is irreducible to the first. Within the context of Leibniz's dynamics, the theory of *vis* thus plays the role of the invariant around which the principle of variation operates. Given a group of transformations for the velocities of masses in a physical system, these variables conserve, within each group, the invariant mv^2.

Second, given the ability to parse the dynamics through the joints of these principles, we can also recognize that the dynamics does not reduce to the theory of *vis* qua mv^2. Anachronistically, we can say that mv^2 can be conserved in a physical system that takes motion to be (ontologically) absolute. Indeed, with different (Newtonian) tools, a demonstration of the conservation of mv^2 can be achieved without appeal to the equivalence of hypotheses. Hence the dynamics can only be understood through a theory of causality based on *vis* rather than merely an explanation of *vis*. In this, the equivalence of hypotheses is an equal participant in the dynamics and indicates why focus should be placed on the causal nature of *vis* rather than on the quantity it represents. To put this in another way, in the language of invariance and variation, we should not privilege either invariance or variation. Rather, we should focus on how the bridge between invariance and variation is constituted. The attempt to understand this bridge is at the core of the dynamics. As we shall see in the next section, Leibniz's misunderstanding of the contribution of the equivalence of hypotheses to physical reality indicates not only his limits as a physicist but also shows the ultimate intentions of his theory. Shedding light on Leibniz's error concerning the identity between the principle of the equivalence of hypotheses and the reduction of all motion to rectilinear motion allows us to understand the central importance of the equivalence of hypotheses to the dynamics and the theory of structural causation that engendered it.

3.3   The Challenge of Rotational Motion

The aim of this section is to analyze the limitations of Leibniz's application of the equivalence of hypotheses. Leibniz incorrectly reduces all motion to rectilinear motion, thereby revealing his presuppositions regarding the nature of causation. This section will also answer the question of the sense in which this theory of motion is tied to his theory of dynamical causation.

As we noted at the beginning of this chapter, Leibniz's relationism about motion can be understood in at least two different ways. There is first the ontological problem of the conditions for the existence of extended motion where what is in question is whether space could exist (and hence whether extended motion could exist) without the existence of bodies. We shall address this with more detail in the sixth chapter. A second way of understanding relationism falls on the epistemic access to the determination and subject of motion. As we have seen, the principle of the equivalence of hypothesis provides a theory of motion based on relative velocities and the same motion, or the same phenomenon, can be given through different variations of velocities in a physical system. Importantly, this view prescribes a limit to our epistemic access to states of motion. As Leibniz remarks, even angels cannot know, for any given physical system, which body is truly moving and which is at rest (GM II 184). This is the general meaning of the equivalence of hypotheses.

If we were to ascribe a scientific meaning to the equivalence of hypotheses, that is, to ascribe the way in which this principle becomes a scientific methodology, we could say that it is the expression of Galilean relativity and the Huygensian method of arbitrary reference (inertial) frames. There is also a metaphysical meaning of the equivalence of hypotheses that renders extended motion itself a reality derivative of the activity of *vires*. As Leibniz explains in the *Discours de métaphysique*,

> [M]otion is not something entirely real, and when several bodies change position among themselves, it is not possible to determine, merely from a consideration of these changes, to which body we should attribute motion or rest. (GP VI, 444; AG 51)

The equivalence of hypotheses thus bars any knowledge of the real subject of motion. In other words, it reveals the distinction between the world of motion and the world of *vires*. From this, we can detect a possible conflict. If the equivalence of hypotheses were simply the expression of methodology, such a principle could sit comfortably alongside motions that do not conform to Galilean relativity. If all motion conforms to the equivalence of hypotheses, however, it must mean that this principle governs over the nature of motion itself rather than simply providing a scientific tool to investigate a category of physical phenomena. In what we will examine below, Galilean relativity and Huygens' arbitrary reference (inertial) frames are limited in their application to rectilinear motion and collision. However for Leibniz, since the equivalence of hypotheses is elevated to an architectonic principle for the dynamics, the problem of rotational motion, which presents a difficulty for the equivalence of hypotheses, threatens the whole edifice of the dynamics. As we shall see, the problem is that the equivalence of hypotheses, as a central component of the dynamics, provides a key aspect to the theory of causation. Thus, Leibniz's use of the equivalence of hypotheses is based on a theory of causation rather than merely a methodology of measurement.

The key here is to understand how the principle of the equivalence of hypotheses relates to a theory of causation rather than merely a method of measurement or determination. For the sake of the generality of the equivalence of hypotheses, Leibniz was forced to argue for the reduction of all motion to rectilinear motion. Our question here is not whether this reduction is warranted. Rather the question is what this reveals about Leibniz's use of the equivalence of hypotheses and the universality of rectilinear motion.

Rotational motion does not provide a challenge to the equivalence of hypotheses because the two are directly inconsistent but rather because it presents a phenomenon of acceleration that is orthogonal to the theory of *vires*. Further, a contextual reading will reveal that Newtonian absoluteness of motion has little to do with Leibnizian *vires* or the problem of conservation in any direct way. This helps us underline the point that, although Leibniz struggled with Newtonian absolute motion over a period of years (1680s–1710s), he did not see it as a challenge either to his own theory of *vis* or, by implication, as a challenge to his own theory of motion. Leibniz saw Newton's position only as a challenge to the *equivalence of hypotheses*. It is this important distinction that I aim to defend here.

The strongest claim to the evidence of absolute motion was put forth by Newton in his famous "bucket experiment" in the *scholium* to the *definitions* of the *Principia Mathematica*.[7] The concave shape formed by a rotating bucket of water cannot be accounted for by the relative motion mechanically imparted by the walls of the bucket. Rather, the concave shape of the water in the rotating bucket is formed by the centrifugal force of the water moving absolutely. For there to be absolute motion, in other words, there must be motion that is not relative to the mechanical imparting of motion from the internal walls of the bucket to the parts of the water in contact with the walls. As the water moves outward from the axis of rotation, it climbs the wall of the bucket and produces the concave shape in the bucket. This concave shape cannot be accounted by this mere contact force and is hence not relative to the motion of the bucket itself (Newton 1972, 10–11).

Another way to understand this "absoluteness" of rotational motion is through its introduction of a break in the symmetry of rectilinear motion governed by the equivalence of hypotheses. The geometry of rotation implies *acceleration* along a curve. This acceleration thus presents a velocity change that is orthogonal to the rectilinear components of the motion considered under an arbitrary (inertial) reference frame (Fig. 3.3).

$$\Delta v = v_1 - v_2$$

---

[7] Isaac Newton, *Mathematical Principles* Vol. I, 10–11. Please note that I will not address the differences between Leibniz and the Newtonians regarding absolute space or space-time. These differences constitute related but nevertheless distinct set of problems. I only note, as Earman has argued, that Newton's bucket experiment does not logically imply absolute space or space-time and that the problem of absolute motion is in principle different from the problem of absolute space or space-time (Earman 1989, 64). I also add, as Stein and Arthur have argued, that the bucket experiment only demonstrates absolute velocity *difference* in rotational motion rather than absolute *velocity* (Stein 1977, 3–49; Arthur 1994, 222).

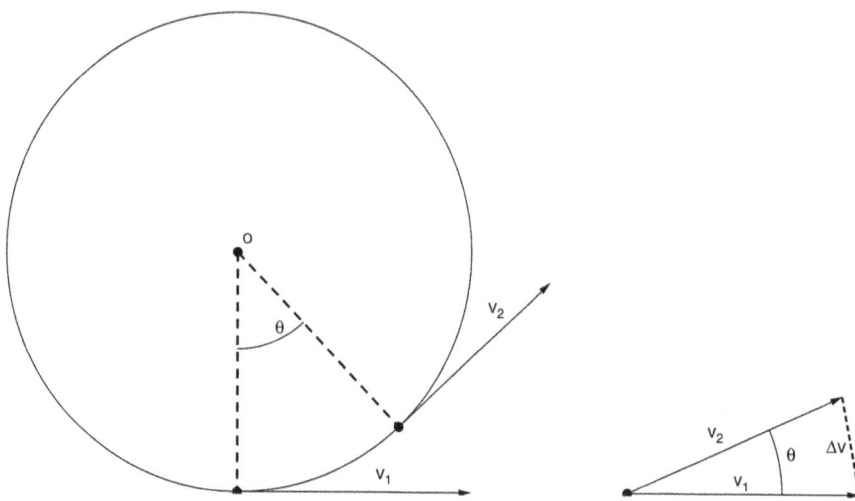

Fig. 3.3 Absolute change in momentum in rotational motion

Rotational motion thus presents a case where velocity change (change in direction) is absolute or, in other words, not relative to the equivalence of hypotheses that governs (rectilinear) inertial frames. This is of course an aspect of the fact that acceleration is "absolute" in this sense for any arbitrary inertial frame. The velocities v_a (of a body a) and v_b (of a body b) are different between inertial frames (where u is the relative difference in velocity):

$$v_a = v_b + u$$

But if we take the time derivative of these velocities (where t is time):

$$(d/dt)v_a = (d/dt)v_b + 0$$

Acceleration or (d/dt)v is thus well-defined given Galilean relativity. Of course, this argument relies on the limited notion of the rectilinear nature of inertia without the later Eulerian concept of angular momentum. As we shall see, Leibniz did see this problem of rotational motion as posing a real threat to his dynamics.

Drawing from work by previous commentators like Stein, Earman, and Bernstein, we can easily point to Leibniz's overall reaction to Newtonian absolute motion, in contrast to Huygens' reaction, during the period of the mid-1690s (Stein 1977; Bernstein 1981, 1984; Earman 1989). Although these commentators point to many interesting turns in this correspondence between Leibniz and his mentor, which resumed after a long break from 1680 to 1688, what concerns us here is what Leibniz's response to Newtonian absolute motion reveals about the Leibnizian construal of the problem of absolute motion (A III 3, 71; A III 4, 368). In the letter of 12/22 June 1694, Leibniz notes that,

> Mr. Newton recognized the equivalence of hypotheses in the case of rectilinear motions; but
> he believes with respect to circular motions, the effort circulating bodies exert to move
> away from the center or from the axis of circulation allows us to recognize their absolute
> motion. But I have reasons that lead me to believe that there are no exceptions to the general
> law of equivalence. (A III, 6, 131)

Leibniz reminded Huygens that in the late 1660s the latter had distinguished
between rectilinear and circular motion, expressing just the kind of difference
between the Galilean relativity of rectilinear motion and centrifugal force in circular
motion that was so neatly captured in Newton's bucket experiment (A III, 6, 131).
Huygens' response on 24 August 1694 acknowledges this former agreement with
Newton but also notes that he had in the last "two or three years" rejected the position (A III, 6, 162). Without entering into the details of Huygens' solution, we note
that he expanded his principal method of Galilean relativity in order to suitably
transform the tangent momentaneous motions of centrifugal rotation to conform to
the method of calculating relative motion prescribed by *equivalence*. As we shall
see below, Leibniz's ultimate solution was not so different from this view, but it
does differ in emphasis from the Huygensian reaction.

The fundamental difference between Huygens and Leibniz on this question of
rotational motion was that Huygens sought to undo the *implication* that the properties
of rotational motion made absolute motion an inevitable conclusion while Leibniz
accepted the implication and sought to recast the problem on two different levels. In
Leibniz's mature *Dynamica* (*circa* 1689–1690), he notes, in the second part, Sect.
3.3, proposition 19 (which we shall return to more intensively in what follows) that,

> I remember [*memini*] a certain illustrious man formerly [*olim*] considered that the seat or
> subject of motion cannot (to be sure) be discerned on the basis of rectilinear motions, but
> that it can on the basis of curvilinear ones, because the things that are truly moved tend to
> recede from the center of their motion. And I acknowledge that these things would be so, if
> there were anything in the nature of a cord or of solidity, and there of circular motion as it
> is commonly conceived. [But] in truth […] it is found that circular motions are nothing but
> compositions of rectilinear ones. (GM VI 508)[8]

This "certain illustrious man" [*viro cuidam praeclaro*] has sometimes been interpreted as referring to Newton. Mormino has recently argued that this certainly cannot be. The "illustrious man" refers in fact to Huygens (Mormino 2011, 701). I
follow Mormino in underlining that "*memini*" refers to a personal memory, a verbal
discussion, rather than referring to something read in book. Further, "*olim*" refers to
a distant past rather than Newton's recent publication of the *Principia Mathematica*.
Finally, Mormino notes that the language of rectilinear and curvilinear motion corresponds more directly with Huygens' language rather than Newton's use of the
"forces receding from the axis of circular motion" in the famous passage concerning
rotation in the scholium following the definitions of the *Principia Mathematica*
(Mormino 2011, 701). What this "memory" refers to is documented in Leibniz's
correspondence with Huygens after the *Dynamica* that we already cited immediately above (A III, 6, 131). While Leibniz was in Paris, Huygens did indeed hold a

---

[8] Translation modified from Stein 1977, 42.

view similar to that of Newton. The view is that curvilinear motion could disclose the "seat" of absolute motion while rectilinear motion, subject to Galilean relativity, could not. As we noted just above, Huygens had changed his mind "two or three years" before his correspondence with Leibniz in 1694. The exact reasons for this change and Huygens' sophisticated development of a physical relativity across rectilinear and curvilinear motion was not known to Leibniz nor to most of the world (perhaps only with the exception of David Gregory in 1693) until the availability of the Codex Hugeniorum 7A (Huygens 1993).[9] Working without a full understanding of his interlocutor's ideas, however, Leibniz was nontheless enthusiastic to find himself united with his former mentor in the defense of the relativity of motion against Newtonianism.

With this complication, we note that, before discussing Leibniz's rejection of curvilinear motions "in nature," we attend to this partial agreement with the idea that rectilinear motion could demonstrate the case for absolute motion. Whereas the late Huygens sought to provide an alternative account whereby curvilinear motion does not have to imply the limited applicability of Galilean relativity, or *equivalence*, to rectilinear motion, Leibniz accepts the implication "in theory". Leibniz's acceptance of this implication, however, indicates that his retort to Newton's challenge makes use of the fallacy of denying the antecedent. That is, if there is "actual" curvilinear motion, then such is a case of absolute-true motion. But there is no "actual" curvilinear motion, hence Leibniz fallaciously concludes that there is no absolute motion, at least not in the Newtonian sense.

It turns out that Leibniz's rejection of curvilinear motion mirrors Huygens' argument in a fundamental way. Instead of working out a version of *equivalence* through the momentaneous tangents that constitute the rotational motion under centrifugal force, Leibniz opts for a decomposition of curvilinear motions into these same momentaneous tangents and, hence, rectilinear motions.[10] In Leibniz's reasoning, since curvilinear motion is reduced to rectilinear motion, curvilinear motions do not really exist in nature but are derivative and hence complex phenomena built up through momentaneous rectilinear ones. Just as Leibniz makes the fallacy of denying the antecedent above, it appears that he might also make the fallacy of composition here. Bracketing these logical problems however, we see that this reductive reading of curvilinear motion is where the similarity between Leibniz and Huygens ends. Compared to the Newtonian account, Leibniz's explanation is a needlessly more complex one since the crucial aspect of the account relies on the mechanical influence of the surrounding parts of the "plenum" that imposes the curvilinear shape on a given sum of minute parts of rectilinear motion.[11] As Leibniz explains in *Specimen Dynamicum*,

---

[9] See Mormino 2011 and Stan 2016a.

[10] See Bertoloni Meli 1990.

[11] Ironically, Leibniz saw his own solution as more parsimonious because he relies on a unique principle of locomotive phenomenon, namely, the principle of *equivalence* (A III 6, 182–283; AG 308).

> For if we assume something we call solid is rotating around its center, its parts will try [*conabuntur*] to fly off on the tangent; indeed, they will actually begin to fly off. But since this mutual separation disturbs the motion of the surrounding bodies, they are repelled back, that is, thrust back together again… (GM VI 252; AG 135–136)[12]

As such, Huygens' approach is certainly the more parsimonious one (in principle) if the aim was to reject the Newtonian challenge to the generality of relative motion. Huygens situates his theory of relational motion through an expansion of the concept and method of reference (inertial) frames, obviating the need to appeal to the plenum.[13]

Despite the obvious shortcomings of Leibniz's method of circumventing Newton's challenge, we see that the former's approach employed a distinction of levels of reality. Newtonian rotational motion is kinematically or phenomenally dissected by Leibniz to constitute a complex motion reducible to something more "real" (rectilinear motion). However it is this "more real" property of motion that provides the criteria for determining any possible form of *phenomenal* motion rather than *vis* or forces. It is crucial to emphasize that Leibniz's strategy here has little to do with absolute motion directly. It is hence a mistake to say that the decomposition of kinematic or phenomenal motion into its "more real" kinematic components reveals "true" motion since the reduction here reveals not true (or absolute) but only relative motion. Again, Leibniz's aim is to show that rotational motion does not escape the physical or phenomenal domain governed by *equivalence*. Both Leibniz and Huygens rely on the reduction of a type of motion that seemingly implies absolute motion to a generalization of relative motion. The difference, again, between Leibniz and Huygens, is that Leibniz's reduction of curvilinear to rectilinear motion situates curvilinear motions at the level of phenomenal reality whereas Huygens accepts curvilinear motions as appropriately real but considers them governed by the same rules as rectilinear motion.

For the sake of completeness, we briefly touch on another treatment of rotational motion in the dynamics project in which Leibniz distinguishes dead from living *vis* (*vis mortua* from *vis viva*). We will address this treatment more rigorously in the next chapter. For the moment, however, the conventional example here is taken from *Specimen Dynamicum*, where Leibniz employs centrifugal force in a rotating body, such as a body in a cylinder rotating about an axis or a sling (GM VI 235–236; AG 121–122). The issue addressed here, as construed by Leibniz, is that the body is "solicited to move" away from the axis of rotation (centrifugal force) but does not move since the body remains tied to a string rotating around the axis or remains within the rotating sling. In such examples as we find in the *Specimen Dynamicum*, all Leibniz aims to establish is the proportional quantitative relation between the solicitation to motion with respect to the rotational axis and actual extended motion. The idea here is that, like gravitational solicitation of a fixed body at some height, there is an intensity (or pressure) within a given body that stands in proportional relation to its possible motion (the fall of a body released from some height). Although he does so erroneously, Leibniz associates centrifugal-centripetal force

---

[12] See Slowik 2006, 624–626.

[13] See Huygens 1993 and Stan 2016a.

Fig. 3.4 Leibniz's
illustration of dead and
living *vires* using rotating
tubes (Figure recreated
from AG 121 and L 438)

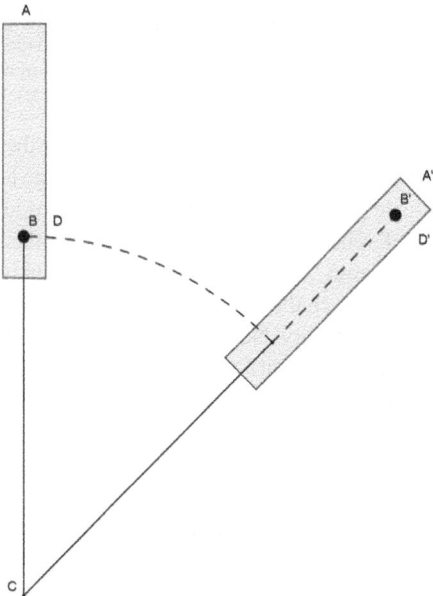

with classical energy dynamics in which the intensity of a potential energy stored up
in a body through rotation or height can be released as kinetic energy. Hence this
extended motion of a body stands in proportional relation to the power [*potentia*]
intensively concentrated in the rotating (but tethered) body. This comparison
between the intensity of potential energy against gravity and the intensity of
centrifugal-centripetal force from the perspective of potential energy is certainly
mistaken, but this may be forgiven since Leibniz, in this context, only sought to
argue for a difference between an infinitesimal magnitude and a finite one governing
the relation between a tendency to move (in the tangent orthogonal to the axis of
rotation) and actual extended (rotational) motion (GM VI 239; AG 121) (Fig. 3.4).

I shall leave the notorious problem of how this infinitesimal-finite proportionality
translates into the physical relation between *vis* mortua and *vis viva* for the following
chapter.[14] Nonetheless, we can remark that Leibniz classifies *vis mortua* (dead force)
as intensity without motion (relative to *equivalence*) and *vis viva* as the expression of
intensity together with motion (also relative to *equivalence*). As such, the intensity-
extensity model here separating *vis mortua* from *vis viva* is a hypothetical one. Thus,
the inherence of *vis* in body is orthogonal to the relation between the translation of
"dead" to "living" *vires*. This distinction between "dead" and "living" in the account
of *vires* has little to do with the *inherence* of *vires* in bodies. Given this independence
it is clear that Leibniz does not consider rotational motion to be a challenge to the
distinction of absolute and relative motion. The Newtonian challenge only pushes
Leibniz to defend the generality of *equivalence*. As such, what is crucial to the distinction between the living and the dead in *vires* is only the problem of the currency

---

[14] See Duchesneau 1994, 223; Gueroult 1934, 38–39.

of a single mathematical framework, grounded in the mathematics of the infinitesimal calculus in the comparison of the momentaneous and extended parts of motion.[15]

In brief, in the *Specimen Dynamicum*, taken as representative of Leibniz's thinking about rotational motion, we find that he treats rotational motion as an instance of the relation between momentaneous motion and extended ones rather than a problem about absolute and relational motion. Hence the various reactions of Leibniz to the strongest (Newtonian) challenge to general relationistic motion resulted in a response that dealt neither with the conservation of *vis viva* as the quantity mv^2 nor directly with the problem of the inherence of *vis* in bodies. Newton's impact on Leibniz, at least on this point, concerns only the range of possible motions: the reduction of complex curvilinear motion to rectilinear ones.

Newton's challenge of rotational motion forced Leibniz to reduce all motion to rectilinear ones. The reason for this is that the universal symmetry of rectilinear motion and collisions provided through the equivalence of hypotheses would be broken by the introduction of rotational motion. To understand why Leibniz was forced to make this error of eliminating the kinematics of rotation through its reduction to rectilinear motion, we make use of the independence of the equivalence of hypotheses and the theory of *vis*.

First, Leibniz's error can be understood through the difference between Huygens and himself when faced with the problem of the bucket experiment. Huygens simply incorporated rotational motion as a species of motion explainable through a composition of rectilinear motions. Leibniz, on the other hand, reduced all motion to rectilinear motion. The difference here is that Huygens *extended* his theory of rectilinear motion to account for different phenomena while Leibniz *reduced* all phenomena to fit the principles of rectilinear motion. In other words, for Leibniz, rotation is a derivative phenomenon. The extension of method, on Huygens' part, and the reduction of the scope of phenomena, on Leibniz's part, are collapsible in terms of the geometrical account given to rotational motion. Their mathematical solutions are similar. They are, however, not collapsible from the perspective of causation. Indeed, this problem of causation was not one of Huygens' concerns. Hence the established methodology could be easily extended to incorporate other phenomena. This is not the case for Leibniz. What was at stake was a theory of causation. In this, Leibniz's plenum theory of the continuity of bodies in space certainly plays a role in providing a convenient supplementary theory for how rectilinear motions became rotational. However, the central motivation remains Leibniz's intent on holding the equivalence of hypotheses as universal. That is, if the equivalence of hypotheses is universal, the reality of rotational motion must be denied since it breaks the symmetry of the relative velocities in rectilinear motion. Of course, we reiterate that this worry itself was an error on Leibniz's part insofar as he lacked the tools to understand angular momentum. For these reasons then, Leibniz did indeed worry that, if the equivalence of hypotheses fails to be universal, then dynamical causation fails to be structural. In other words, the translation of *vis* to motion could pass through a principle other than the equivalence of hypotheses.

---

[15] We shall examine some mathematical aspects of this in the next chapter.

Second, given the independence of the principle of *equivalence* from the theory of *vires*, we can grasp the logical distinction between inertial motion and the conservation of *vis* qua mv^2. However, the theory of structural causation requires that this conservation be expressed through a phenomenal framework defined by the equivalence of hypotheses. Leibniz's error is thus his failure to see how the theory of *vis* qua mv^2 could be conserved in cases of rotational motion. This error demonstrates Leibniz's implicit commitment to a theory of structural causation where what causes is *vis* and what is caused are phenomena. The bridge between cause and effect is most importantly operated through the equivalence of hypotheses. It is for this reason that the equivalence of hypotheses is universal and all phenomena must reflect this structure as the form of phenomena. If the role played by the equivalence of hypotheses in dynamics were merely that of a methodology for using arbitrary inertial frames, Leibniz's reduction of all phenomena to motions conforming to rectilinear motion would certainly be excessive. However we can appreciate why Leibniz argues for this reduction if we understand that the form of physical phenomena is the means through which dynamic cause, *vis*, becomes locomotive effect, phenomenon.

3.4   Inertia and *Vires*

In order to provide a fuller treatment of how the equivalence of hypotheses provides a theory of the form of locomotive phenomena, it is necessary to examine the relation between the equivalence of hypotheses and Leibniz's concept of natural inertia [*inertiam naturalem*] (GP IV, 510; AG 161). The aim here is to concretize the distinction between causation through *vis* and the form of phenomena.

As the term *vis insita* in the Keplerian tradition indicates, the literal sense of this term comes closest to the notion of an inherence of "force" in bodies (GP VII, 313). This is, however, quite misleading. The Keplerian notion of inertia should be understood as the "force" *of* bodies, or better yet, of *matter*, understood as the "passive" metaphysical constituent of a body. This "force" of matter then would be the "passive" contribution of matter to the "active" properties of corporeal motion. From the Neo-Aristotlelian tradition this "natural inertia" was primarily a tendency of a massive body to decelerate motion *qua* "*inclinatio ad quietem*" (Kepler III 1990, 256; 1992, 407).

Although Leibniz's conception of *vis insita* does indeed correspond to the idea of a "force" of matter, it is difficult to provide a systematic link between this notion to that of inertia in classical mechanics. Without providing a comprehensive treatment, we look at one single (but central) text where Leibniz uses the concept in two different ways: the 1698 *De Ipsa Natura* (GP IV 504–516; AG 155–167). After having commended Kepler on the use of the term "natural inertia," defined as the resistance of mass to motion in the vein of *inclinatio ad quietem*, Leibniz adds that, "And just as there is natural inertia opposed to motion in matter, so too in body itself, indeed in all substance, there is a natural constancy opposed to change" (GP IV 510; AG 161). In one and same place, Leibniz invokes inertia through the Keplerian *inclinatio ad quietem* and associates it with a "natural constancy". The ambiguity of

Leibniz's use of "natural inertia" and its associations in this and other texts makes it hard to concretely grasp his notion of inertia. However, I follow commentators such as Bernstein and Fichant in holding that the Galilean and Cartesian view of inertia as the persistence of a state (of rest or motion) was one of the earliest pillars of Leibniz's understanding of the laws of motion, harking back to his earliest work on mechanics.[16] Hence, even if Leibniz was in this late work of 1698 still full of praise for Kepler, he actually rejected the notion of an *inclinatio ad quietem*.[17] More importantly, Leibniz demonstrates a certain comprehension of classical inertia when, in the lines after the above quoted, he states that,

> [A]s certain as it is that matter cannot initiate motion through itself, it is just as certain that a body conceived in and of itself retains an impetus once it is imparted, and remains constant in its mobility [*levitas*], that is, it has the tendency to persevere in that series of its changes... (GP IV 511; AG 162)

It is important to note that what "perseveres" in this case is impetus, which Leibniz defines through the Cartesian quantity of motion (product of mass and speed). As such, one might easily associate the classical concept of inertia (conservation of momentum) with this "tendency to persevere" here, though Leibniz himself does not explicitly make this connection. Bernstein also emphasizes that this Leibnizian "tendency" does not become conceptually "inertial" (in the classical sense) except when the motion is eventually constrained in some way, such as in collision, the resistance of a medium, or other cases of momentum change (Bernstein 1981, 107–108). This is stated later in the same text when Leibniz notes, "[J]ust as force is necessary for producing motion, so too, once an impetus is given, far from requiring a new force for continuing motion, one needs a force to stop it" (GP IV 515; AG 165). It turns out then that Leibniz's idiosyncratic use of the proto-inertia concept reduces neither to the Keplerian one nor strictly to the classical-Newtonian one. In the case of unconstrained uniform motion, what we do see explicitly connected is the impetus and its "perseverance" in motion, or a "series of its changes".

In this example of unimpeded or inertial motion, the situation here is "forceless" (in the Newtonian sense), as Bernstein puts it (Bernstein 1981, 108). But it deals with neither Newtonian force (since there is no momentum change) nor Leibnizian *vis* (since although energy-work is conserved, it is not at stake here). As Leibniz himself insists, "conservation by a universal cause necessary to things is not at issue here..." (GP IV 515; AG 165). Since the issue of *vis* qua conservation or universal cause is not at issue, from the perspective of *vis insita*, a body at rest and a body in uniform rectilinear motion are equivalent. Thus, we can equally hypothesize both with respect to any given body. What is conserved in the case of unconstrained rectilinear motion is *impetus*, the Cartesian quantity of motion, *vis insita* reduces to inertial motion governed by *equivalence*.

---

[16] See Bernstein 1981, 101 and Fichant 1995, 73–76.

[17] Leibniz most explicitly rejects Keplerian inertia qua *inclination ad quietem* while recognizing Kepler's contribution to the problem in his 24 March/3 April 1699 letter to De Volder (LdV 72–73). Here, Leibniz further argues that inertia applies only to rectilinear motion, thus revising the Cartesian (and Spinozist) adoption of the principle "*quantum in se est*".

Thinking through the *equivalence* principle, Leibnizian "inertia" is another perspective with which to understand the independence of *equivalence* and the theory of *vis*. We see that Leibnizian "inertia" is equivalent to the principle of *equivalence* and only becomes "inertial" when there is collision or some other form of resistance at stake. This is not only a version of Leibniz's rejection of Cartesian mechanics *vis-à-vis* the inadequacy of the quantity of motion but also an indication of the distance that separates classical mechanics and Leibniz's theory of *vis*.

From this we can understand that the equivalence of hypotheses "stands in" for the proto-inertia concept. Again, the reason for hesitating to grant Leibniz a fully developed theory of inertia is because its systematic relation to momentum change was not well-established. The strict implication here is that the tendency of the perseverance of impetus applies only to the structure of phenomena. This renders this notion an instance of the equivalence of hypotheses. We shall deepen this view by examining an important section of the *Dynamica* that has attracted much attention in its reception over the years.

In the Proposition 19 of the third section of the second part of the *Dynamica*, Leibniz argues for a slightly modified version of the equivalence of hypotheses:

> Proposition 19. The Law of Nature that we have established of the equipollence of hypotheses –that a Hypothesis once corresponding to the present phenomena will then always correspond to subsequent phenomena –is true not only in rectilinear motions (as we have already shown), but universally: no matter how the bodies act among themselves; but provided that the system of bodies does not communicate with others, i.e., that no external agent supervenes. (GM VI 507; Stein 1977, 41)

What is striking here is that Leibniz's statement of *equivalence* is transformed into a diachronic principle. The traditional statement of *equivalence* provides for the synchronic variability of assignations of velocities to bodies. Here, this novel expression of the "equipollence" of hypotheses in the *Dynamica* posits that each hypothesis (each set of variable distribution in a system) continues to hold for the temporal evolution of the system. With this shift from synchronic to diachronic, the term "equipollence" is an apt term because it aims to establish the idea that, within each hypothesis, a certain distribution of values, the same measure of *potentia* unfolds in time. This temporal unfolding of *potentia* through *actio* has been examined in the previous chapter. Here, the aim of the proposition is to state the universality of the equipollence of hypotheses with respect to the range of motions (rectilinear and otherwise).

The aim of this proposition has often been misunderstood due to the fact that it is ambiguous what is in fact "universal" in the proposition. "*Non tantum in motibus rectilineis (ut hactenus ostendimus) sed et in universum vera est, quam stabilivismus Naturae Lex de aequipollentia hypothesium...*" (GM VI 507). Universality could apply to the equipollence itself, the measure of *potentia* across time and across hypotheses, or it could apply to the range of motions (rectilinear and otherwise) that fall under the principle. Since the *Dynamica* spends the previous propositions defending the equivalence of hypotheses as such, it appears that the "universality" in question here concerns the universal application of the principle across a range of different kinds of motion.[18] This contextual reading contributes to the understanding that the principle of *equivalence* and, here, the *equipollence* of hypotheses, is universal across rectilinear, rotational and other forms of motion.[19] The widening scope of the principle of equivalence is hence generalized, in the form of the equipollence of hypotheses, to the whole range of possible motion. As we saw in the previous section, the concrete meaning of this is the reduction of all locomotive phenomena to rectilinear motion.

Again, here we note that the insufficient formation of the concept of inertia forces Leibniz's hand to reduce all motion to compositions of rectilinear motion. From the perspective of a retrospective evaluation of this from classical mechanics, this Leibnizian concept of "inertia" is insufficient precisely because it erroneously requires the reduction of all motion to rectilinear ones, that is, it requires the reduction of all motion to the rectilinear framework required by the equivalence of hypotheses. In this, the "universality" of equipollence of hypotheses shows Leibniz's commitment to a theory of the form of phenomena distinct from any commitment to the theory of *vires*.

The equipollence of hypotheses nonetheless states the conservation of *potentia* across different hypotheses, a distribution of variables in physical systems that evolve in time. We have already seen in the last chapter how Leibniz's theory of *actio*, or the unfolding of *vis* qua cause in time, is the principle that guarantees this conservation: $a/t = ms/t \cdot s/t = mv^2$. We also examined how action ranges over inertial (or "forceless") motion and motions involving acceleration. For the first case, velocity simply determines the distance traveled over time, as duration increases distance (the variable s) increases proportionally. This reproduces the conservation of mv^2. For the second case, action allows Leibniz to divide his calculation of mv^2 into two factors of velocity. Hence, the change of velocity will be inversely proportional to the change of the formal effect of the system.

$$a = ms \cdot v$$

$$a / \Delta v \propto \Delta ms$$

In short, the theory of *vires* is capable of providing an account of the equipollence of hypotheses in a way that does not rely on the equivalence of hypotheses. We take this to imply, again, the independence of the two theoretical aspects of Leibniz's dynamics. More importantly, we have developed here a deeper account of the scope of Leibniz's equivalence of hypotheses. Namely, we have shown that the principle extends beyond the proto-inertia concept of Leibniz's inherited Huygensian

---

[18] The translation cited above favors my interpretation.

[19] *Aequipollens* and *aequivalens* are used rather interchangably throughout Leibniz's mathematical and physical writings. Even in the dynamical writings Leibniz will use the expression "quod effectus integer semper aequivalere debet causae suae plenae" (Leibniz 1991, 454; 2007, 696–697). In mathematical treatises, Leibniz will say, "Ego vero cum Theorema quoddam generalissimum reperissem, cujus ope quaelibet figura in aliam plane diversam sed dimensione aequipollentem converti potest…" (A VII 6510). However, there are reasons to see that, within the confines of this context, *aequipollens* means the equality of "power" across the range of motions.

methodology where it was effective in establishing the basis for the symmetry of collision and the measure of mv^2. Disassociating the equivalence of hypotheses from inertia also allows us to see that the principle, far from being related to momentum change, has been elevated into a principle that provides the form of locomotive phenomena as such. Given Leibniz's improperly formed concept of inertia, this pushes Leibniz to reduce all motion to rectilinear ones.

3.5   The Equivalence of Hypotheses and Dynamical Causation

In the previous sections of this chapter, we examined the independence of the equivalence of hypotheses from the theory of *vires*. In this section we will attempt to provide an account of how this independence of the two theories will allow us to understand dynamical causation through a synthesis of the two principles. Here, the aim of the chapter should already be clear enough. In general, through the principle of equivalence, Leibniz reduces all motion to rectilinear ones and hence benefits from the symmetrical structure well-defined through the methodology inherited from Galilean relativity and Huygensian arbitrary (inertial) reference frames. The way in which Leibniz goes about establishing a reduction of all motion to rectilinear ones has been examined through its multiple limitations and errors. These errors demonstrate that the equivalence of hypotheses constituted more than a methodology but also provided a quasi-metaphysical or, to speak anachronistically, transcendental theory of the form of locomotive phenomena. This means that *vis*, or the cause of motion, can only be expressed as motion through a form governed by the principle of equivalence. We also saw that the theory of *vires* is independent of the development of the principle of equivalence. Hence regardless of which quantity is actually conserved in nature, the principle of equivalence holds. The result here, the aim of this chapter, is to show that dynamical causality is one where Leibnizian *vires* are expressed through physical systems, physical phenomena, for which the principle of equivalence holds. Hence, dynamical causation is structural. The causal relation holds only between the action of *vis* and a system of bodies whose motion is relational.

Hence the first component of structural causation, the measure of *vis*, mv^2, is a structural property of a physical system. When considering a one-body system, that single body is understood, through its motion, as expressing the quantity mv^2. When considering a many-bodied system, mv^2 is the sum measure of all the motions of the all the bodies together. This is simply the meaning of a conserved quantity. As we saw, what makes *vis* more than a property is the concept of action. If *vis* were merely a property, we would face the problem of the inactuality of *vis* in motion since the *vis* could only be expressed by the capacity for *future* motion. If this were the only way through which *vis* could exist, it would be inactual, constantly separated between the accomplishment of past effect and the *potentia* to produce future effect. Action solves this problem by providing the constant translation between *potentia* and effect. This concept allows us to maintain the actuality of *vis* through its constant action in the temporal unfolding of *potentia* and effect. Causation is structural in this sense because *actio* is the translation of cause qua *vis* into phenomena unfolding in space and time.

The second sense of dynamical causation, examined in this chapter, is that locomotive effect can only be considered as many sets of motions mapped onto each other as a group. Relatively defined velocities, considered as a set of variables for the bodies of a physical system, can be continuously mapped onto one another. This principle, the equivalence of hypotheses, is treated by Leibniz as a fundamental feature of phenomena, and all seeming exceptions can be resolved by reducing them to rectilinear motions governed by this same principle. This is thus the second way that dynamical causation is structural. *Vis* qua cause acts on a potentially infinite set (forming a group transformation) of motions. The reason for this second sense is different from the first sense. The reason here is that motion qua phenomenon itself is governed by the equivalence of hypotheses. Hence due to the theoretical independence of the theory of *vires* from the theory of phenomena, even if the theory of *vires* were false, all motion would still fall under this group structure. But if the theory of *vires* is true, this multiplicity of phenomenal effects would be provided with an invariance, mv^2, around which relative motion varies. The causal connection between *vires* and motion would then be structural in the sense that cause and effect would be a relation across ontological levels. As such, phenomenal motion can only be related to its cause through the systematic distribution of the invariant quantity mv^2 across time and across space. In other words, the equivalence of hypotheses provides a key element in the understanding of dynamical causation as structural because it provides the principle through which *vis* is translated into a group of internally relative variables. *Vis* is thus causal structurally because it is both expressed and limited by the structure of phenomena. In other words, since no motion can be expressed outside of the structure of the equivalence of hypotheses, *vires* can only produce effects of this form. This form of phenomena naturally makes dynamical causation structural. The causal factor is *vis*, and what is caused is a group of variations answering to the form of phenomena.

It is from the point of view of this synthesis that the theory of causality in the dynamics is to be understood as structural. When the causal relation is that of *vis* to a locomotive phenomenon expressible through a group of internally related positions and speeds, we have truly left the domain of mechanical explanation. In fact, the equivalence of hypotheses renders it impossible to place *vires* and motions on the same level of reality precisely because the same locomotive event is expressible in a potentially infinite set of variable distributions. Instead of basing a theory of causation on mechanical relations, what we have instead is a concept of causation based on the production of phenomena through non-phenomenal *vires*. The role of the equivalence of hypotheses is thus to provide the delimited form that any motion must take and hence to serve as the ontological barrier that formally or ontologically separates motion from *vires*, or effect and cause. At the same time of course, the equivalence of hypotheses also serves as the mode through which *vires* are translated into motion. As we have discussed, given merely the principle of *equivalence*, we do not have sufficient tools for discovering and demonstrating the spatial-temporal *structure* of motion and the invariants that govern it. The demonstration of the conservation of mv² and the distribution of velocity in the evolution of a physical system requires an independent theory of *vires*.

3.6 Concluding Remarks

Leibniz frequently emphasizes the irreducibility of his dynamics to the geometry of motion.

> [B]ecause we cannot derive all truths concerning corporeal things from logical and geo-
> metrical axioms alone, that is, from large and small, whole and part, shape and position...
> we must admit something metaphysical, something perceptible by the mind alone over and
> above that which is purely mathematical... we must add to material mass [massa] a certain
> superior and, so to speak, formal principle. (GM VI 241; AG 125)

A theory of motion based on cause qua *vis*, the dynamics, represents a *nova scientia* precisely because the object of the dynamics is *vis* and its derived effects in space and time. The reason for this is because, due to the equivalence of hypotheses, the relative character of phenomenon bars any direct knowledge of *vis*. This epistemological limitation of geometrical measure is the Keplerian inheritance in the dynamics. Nonetheless, the dynamics renders this negative limit of epistemic access into a positive principle and method of his dynamics. The equivalence of hypotheses, understood through Galilean relativity and Huygensian methodology, becomes a way to understand the mode through which non-phenomenal *vires* are expressed in phenomena. A limitation becomes a positive methodology for constituting the form of phenomena.

In this chapter, we have examined this form of phenomena through the many missteps and limitations of Leibniz's theory and its development. Leibniz certainly suffers from both logical errors (affirming the antecedent, fallacy of composition) and methodological limitations (incomplete concept of inertia). But through examining these missteps, we have revealed Leibniz's systematic intention to fulfill a theory of causation. Understanding the principle of equivalence provides us with a way to grasp this conception of causation more concretely.

Throughout the dynamics, Leibniz pursues a theory of causation that never denies the applicability of efficient causation to explain motion. Nonetheless, Leibniz's dynamics would be superfluous if purely mechanical relations were sufficient to account for motion. Leibniz often has to resort to the ambiguous language of "higher" and "lower" to present the convergence of these different modes of causation. "[W]e acknowledge that all corporeal phenomena can be derived from efficient and mechanical causes, but we understand that these very mechanical laws as a whole are derived from higher reasons. And so we use this higher efficient cause only in establishing general and distant principles" (GM VI 242, AG 126). This "higher efficient cause" [*causa efficiente altiore*] no doubt refers to the causation provided by the *actio* of *vis*. However, at such great "height" this cause stops being "efficient" in the common sense of the word. It is at the clarification of this kind of division of causes into "higher" and lower," or "primary" and "derivative," that our examination of the principle of equivalence aims.

Given the reduction of all phenomena to rectilinear motions and collisions universally governed by the equivalence of hypotheses, we grasp why the dynamics provides room for both efficient causation and "higher" causes. Through the reduction of all motion to rectilinear ones, all phenomena and hence all efficient causation and mechanical relations exist in the form prescribed by the principle of equivalence. It presents a closed structure of distributions of variables and positions. Nonetheless, *vires* act through this delimitation of phenomena. The *actio* of *vires* is thus imperceptible merely from mechanical relations restricted in the way Leibniz understood them (restricted to rectilinear relations). Hence the theory of *vires* constitutes a "higher" and more "primary" form of reality that is ultimately responsible for the structure of locomotive phenomena. The theoretical independence of the principle of equivalence and the theory of *vires* thus grants us the needed distinction between two different levels of reality while granting primary causation to *vis* rather than the efficient causes occurring through mechanical relations.

Hence, efficient causation must be understood as derivative to dynamical causation. The theory of causation from the perspective of the dynamics establishes a relation between *vis* and the whole range of efficient or mechanical relations operating among phenomena. Dynamical causation operates the relation between *vis* and this range of mechanical relations rather than being one among these mechanical relations. Dynamical causation is thus structural insofar as what causes and what is caused constitute different levels of reality. In addition, the *actio* of *vis* produces an effect which can only, through the principle of equivalence, be expressed as a group of mutually relative variables. The structure of these variations around the invariant mv^2 is thus an apt representation of how *vires* is expressed in motion but irreducible to locomotive phenomena itself.

Chapter 4
Continuity and Causation in the Dynamics

Abstract This chapter continues with a three-part presentation of the central architectonic components of the dynamics. In this chapter, we examine the status of continuity in the theory of motion developed in Leibniz's dynamics. The chapter traces some difficulties in the development of continuous motion in order to highlight the productive frictions between Leibniz's attempts to provide a geometrical and dynamical account of motion. The status of continuity will highlight what is at stake in Leibniz's focus on *vis* as the *cause* of physical phenomenon: the distinction as well as the connection between ideal (geometrical) continuities and dynamical action.

4.1 Introduction

The continuity of motion is easily taken for granted. Continuity seems to be a brute or quasi-brute quality of motion arising from the continuity of space and time. This was certainly not taken for granted by Leibniz, who dedicated a number of earlier works like the *Theoria motus abstracti* (1671), and later the *Pacidius philalethes* dialogue (1676), to giving an account of the continuity of motion. Although Leibniz never provided a final or satisfying answer to this question in the early period, it was clear that a more mature Leibniz used the continuity of motion as the criterion for qualifying its "mere" phenomenal nature. The continuity of motion entails, in Leibniz's mature physical theory, its indeterminateness. This was why a concept of *vis* was required to guarantee the identity and formal determinateness of motion. As we have argued in the last chapter, the equivalence of hypotheses constitutes a principle in the dynamics as well as a methodological pillar of the project. From this perspective, the role played by continuity is to serve as the condition for the variation of motion produced by *vis* as its acts in space and time. This variability or "relativity" relies on the continuity of the variables in the range of space and time. Although we have examined the equivalence of hypotheses in the previous chapter, through the concept of the continuity of *motive effect*, the crucial question of the continuity of how *vis* acts in space and time such as to produce this continuity is unaddressed.

Before sketching out my argument, let me clarify what I am not arguing. I hold that there is no direct relationship between the doctrine of *vis* qua cause of motion and the property of continuity that results. Of course Leibniz's equivalence of hypotheses requires that motion is continuous as an assumption. We can thus

© Springer International Publishing AG 2017
T. Tho, *Vis Vim Vi: Declinations of Force in Leibniz's Dynamics*, Studies in History and Philosophy of Science 46, DOI 10.1007/978-3-319-59055-4_4

approach the question of the continuity of motion in a negative way. Can motion be continuous if the invariance of *vis* did not hold for physical phenomena? Since the continuity of physical motion can be reducible to the continuity of abstract geometrical motion, motion certainly can be continuous without the invariance of action. This would simply follow from the fact that motion is spatial displacement and since space is continuous, motion is also continuous. Of course, motion could also be the discrete translation of a body across discrete contiguous parts of space. Leibniz in fact held this view for a period of time.[1] However, nothing about the conservation of *vis* allows us to decide between these two options.

So what does the theory of *vis* add to the question of the continuity of motion? There is no direct relationship between *vis* and continuity, but there is an indirect one. This indirect relationship is due to the role of determination that *vis* plays in Leibniz's theory of corporeal motion. Since the reality of motion is grounded on the reality of *vires*, with the emergence of the dynamics, Leibniz is released from the burden of having to provide an account of the consistency and determination of motion from the perspective of its geometrical composition. As such, an account of how the continuity of motion is composed can be referred to a purely mathematical or abstract problem answered by the mature theory of the foundations of the infinitesimal calculus and geometry.

Let me then lay out the aims of this chapter. Leibniz's mature view of continuity is that whatever appears as continuous (extended bodies, motions, transformations of other sorts) are only results of more ontologically fundamental discrete entities or contiguous aggregates of entities. This view is composed of two aspects. First, continuous extended wholes can be subdivided to infinity without attaining a final indivisible part. Wholes are defined as continuous just in the case where continual subdivision is indefinite. Second, this continuity qualifies these infinitesimal parts of extended wholes as not entirely real or only virtually real. Only what is discrete is real. Now, what this view entails is somewhat counter-intuitive. A whole is real because it is a discrete whole. However its indefinite parts, produced continually by subdivision, are not real because they only gain a degree of reality by virtue of being parts of a discrete whole. Conversely, points are real because they are discrete divisions between segments of continua. Points mark the extrema of bounded extended line segments. Since distinct points can only mark analytic moments on a pre-given continuous extension, points cannot make up a continuum.

From these two aspects, two points also follow with respect to motion. First, the continuity of motion cannot be composed out of anything other than smaller and smaller parts of continuous motion. Secondly, since the continuity of motion is only composed out of other continuities, motion is only real by virtue of some ground that provides a criterion of unity. The reality of motion must be grounded in a dis-

---

[1]Leibniz proposed a model for understanding motion through the discrete translation of a body through contiguous spaces in the 1676 dialogue "Pacidius Philalethes" (LC 127–221). However, in Leibniz's mature period, he also held a different view that there are "real" divisions in motion, rendering motion and other transformations "not really continuous". The language of "not really" is to be understood in context. This more complex view of the mereology of motion will not be addressed here (LdV 327; GP II 278–279). See also Leibniz to Electress Sophie 31 October 1705 (GP VII 563; A I 25, 202).

crete whole, a role played by *vis* in the dynamics, that guarantees the identity of "a" motion. Moreover, analytical points within continuous motion are real by virtue of their discreteness but cannot serve to compose a motion, just as points cannot compose a line. As Leibniz puts it in the *Dynamica*, we might say that, just as points *constitute* but do not *compose* a line (since points fall on a line), there are moments of motion that constitute an extended motion that do not compose the motion: "*constitui dico, non componi*" (GM VI 370). In this sense, we can argue that the theory of *vis* allows us to provide a theory of the determinateness of motion without an account of how the continuity of motion is composed.

The aim of my argument in this chapter will thus be to establish the following thesis. The continuity of motion is a necessary feature of the phenomena of motion independent of the doctrine of *vis* qua cause of motion. However, this continuity can be understood as *necessitate ex alterius hypothesi* because it is through the determination of a more fundamental structural foundation for the reality and actuality of motion that releases Leibniz from the burden of addressing continuity through the composition of the parts of motion. It thus follows from this argument that the relationship between cause and effect within the dynamics should be interpreted through a structural division between a domain of continuous extended motion and a domain of non-extended *vires* where the latter is the cause of the former.

This chapter will proceed in four sections. In the following second section, I will provide some context for understanding the role that continuity plays in Leibniz's thinking about the nature of motion. Here we turn to some of his earliest texts in order to illustrate the difference in strategy that he pursued in his earlier period and his post-Paris (post-1676) period. In his earlier writings, Leibniz saw geometrical and physical continuity as different in kind and sought to ground a theory of physical motion from the perspective of the composition of physical motion from a problematic theory of unextended motive tendencies. In the third section, I will move from this context to examine how Leibniz, in his later period, collapsed the distinction between abstract geometrical continuity and the continuity of actual physical motion. I argue that the role of grounding motion shifts from geometrical composition to a metaphysical appeal to the individuation granted by the concept of *vis*. This implies two related claims. It implies that the geometrical composition of motion can only be reduced to a relation between continuous parts of motion. It also implies that continuity of motion is an irreducible feature of locomotive phenomenon. From this, I move to the fourth section, where I examine the famous "missing" factor of ½ from the quantity mv^2. This problem has often been seen as a mistake on Leibniz's part. I argue that, although this manner of treating the conservation quantity demonstrates a certain limitation in Leibniz's method, it does not indicate an error. This reinforces the claim that the motion caused by *vis* qua mv^2 cannot be reduced to discrete momentaneous parts represented by *vis mortua*. Finally, I will connect the idea of the necessity of the continuity of motion *ex hypothesi* with the concept of a causal principle that moves across ontological lines. This renders the best way to understand how the indeterminate and imaginary nature of continuity can follow from the determination of *vis*. In closing, I leave the reader with a speculative idea that will be picked up in the next chapter.

4.2   Phoronomy and the Continuity of Motion

It is perhaps odd, in the first place, to ask why motion is continuous. Since motion is the change of place of a body in time, the continuity of motion, as we have mentioned, seems to follow quite naturally from the continuity of abstract space and time. The seeming bruteness of this feature of physical motion was certainly not taken for granted by Leibniz. In Leibniz's early work on physical theory leading up to the dynamics, the continuity of motion was put into question a number different times from a phoronomic perspective.

Now, phoronomy (*phoronomia* or *phoranomica*) refers to the geometrical analysis of motion. More specifically, it refers to the geometrical analysis of the path traced by corporeal motion, a method developed by Joachim Jungius in his *Phoronomica sive doctrine de motu locali*.[2] The basic idea here, from the Greek root φέρω (*phero*), is to treat laws of motion from the perspective bodies "bearing" or "carrying" themselves through an extension. It can be understood as a forebear to what we now call kinematics insofar as both are concerned primarily with the geometry of motion rather than its causes (dynamics). However, since kinematics as we now understand it is defined in a post-Newtonian manner through the analysis of the path of motion, this anachronism may lead to more misunderstandings than clarification.

In its original context, phoronomy relies on an *a priori* geometrical analysis of motion, relying much more on the tools of geometry itself than the analysis of the "forces".[3] In the post-Paris period (1676), he will also remark that, "Geometry is Mathematical Logic, so I will boldly declare that Phoronomy is Physical Logic" (A VI 3, 533; LC 137). The idea here is that a "logic" establishes the "science of gen-

---

[2] The term *phoranomica* and its cognates, drawn from Jungius' text, is part of Leibniz's lexicon for discussing laws of motion since the earliest years for example, in the 1671 *Theoria motus abstracti*, and will serve as the title for his 1689 treatise *Phoranomus seu potentia et legibus naturae* (GM VI 79). Jungius dies in 1657, and the text in question was not published in his lifetime. It is equally uncertain when Leibniz first reads the treatise. It is clear that Leibniz held Jungius in very high regard, judging him more profound than Descartes (who apparently was unfortunately detained in the "antechamber of true philosophy") in a letter to Christian Philipp (A II 1, 767). Leibniz also refers, in various writings, to Jungius' *Logica hamburgensis* (1638) and *Geometria empirica* (1627). Robinet suggests that Leibniz might have read Jungius' *Phoronomica* in 1679 and certainly in 1689, evidenced by a copy transcribed in his own handwriting (Leibniz 1991, 532). Leibniz's correspondences in 1686 show a number of letters discussing the publication of Jungius' diverse *Nachlaß* including the *Phoronomica* (A II, 2, N. 20–23). See also Duchesneau 1994, fn. 166.

[3] Jungius' *Phoronomica* prescribes a method that alternates between geometrical and physical (*qua* phoronomical) propositions in order to map geometrical relations onto the spatial figures traced by locomotion. Theorems in the treatise are deduced through the logical synthesis between geometrical definitions and the extensions traced by motion (Jungius 1699, 3–9). Regardless of how one is to evaluate Jungius' treatise, Leibniz himself saw it as limited to *a priori* or "purely" geometrical methods (with logic) instead of availing itself of the concept of *vis*. Leibniz will insist on this limitation of Jungius in the 1689 *Phoranomus sive potentia et legibus naturae* where he argues that Jungius' method is insufficient insofar as it fails incorporate *vires* (Leibniz 1991, 483, 2007b, 754–755).

eral reasons" in that domain (A VI 3, 532; LC 137). In a late period, Leibniz will state that, "Phoronomy [*Phoranomica*] is two-fold, one without the consideration of time… another that involves time, such as the tracing of accelerated or similar motion. Such as [that which] composes motion in a geometrical and physical way" (C 525–526).[4] The mature Leibniz understands the phoronomic method as a "purely geometric" analysis of motion through the extension that a moving body traces. As he argues in *Phoranomus seu de potentia et legibus naturae*,

> Phoronomics [*Phoranomices*] is thus named not without intelligence by Jungius in his book of this title, where it simply investigates the lines traced by motion as part of the doctrine of motion; this is purely geometrical. However, the laws of nature concerning the communication of motion is to be observed by motive force [*vires motrices*]. (Leibniz 1991, 483, 2007b, 754–755)[5]

The prior notion of phoronomy, as we will examine, is a necessary but insufficient basis for the dynamics.

Despite the limitations of phoronomy, how does this help us understand motion? In the first place, it allows us the tools to analyze and determine motion's continuity. One of many possible ways to interpret this is to see the phoronomic method as a means to handle the Eleatic paradoxes. Recall that Eleatic paradoxes were essentially generated out of the mapping the continuous onto the discrete.[6] The Eleatic paradoxes result in contradiction when the parts of the continuum end up being greater than the whole or when the whole is quantitatively unresolvable into discrete parts. This only results, however, if we identify the parts of the continuum with a sum of discrete unities. As Leibniz understood it, the famous "labyrinth of the continuum" is the result of confounding continuity and discreteness (LdV 333). Yet without confounding the two, we can nonetheless establish some kind of relation between the two that does not generate the paradox. Indeed, to grant determination to continuous motion, this relation between continuity and discreteness must be established. This is then what Leibniz aims to do in his many attempts at providing a phoronomy.

---

[4] For historical contrast, we see that phoronomy has taken up another meaning by the first decades of the eighteenth century. In J.H. Zedler's *Grosses vollständiges Universal-Lexicon aller Wissenschafften und Künste* (1731–1754), the term is understood as "the science of the motion of solid and fluid bodies" and includes "mechanics, statics, hydraulics and aerometry in it" (Zedler 1731–1754, 2193). The term is strictly associated with Jakob Hermann's 1716 *Phoronomia, sive de viribus et motibus corporum* with mention to Newton, Leibniz, and Bernoulli. Jungius is not mentioned. In D'Alembert's 1785 *Encyclopédie méthodique*, phoronomy is the "science of the laws of equilibrium, the motion of solids and fluids" and is represented (in D'Alembert) exclusively by Jakob Hermann's 1716 *Phoronomia* (without mention of Jungius or Leibniz). Given that Leibniz himself felt free to dissociate the term "*phoronomia*" from its original Jungian context, it is no surprise that the term could receive other meanings in the following decades. Though Hermann is a later contemporary of Leibniz, the context for physics and natural science as a whole had shifted. In any case, this term is used differently in Leibniz and his interlocutors in the late 1660s and early 1670s (D'Alembert 1785, 577).

[5] See Duchesneau 1994, 165–166.

[6] The fletcher's paradox is not directly obvious as a case of mapping continuity on the discrete, but it plays on the same irreducibility of continuous motion to discrete moments.

In the early physical work of Leibniz, a phoronomy was developed with an eye to show how to treat continuity while avoiding the Eleatic paradoxes. As such, a theory of motion must be logically built from carefully ordered definitions that avoid the logical snares of paradox. This approach, of course, blurs the lines between geometry and physics as we are now accustomed to them. For instance, when contemporary geometers talk about the "speed" of a curve, they have in mind the growth of one set of variables with respect to the growth of another set. Though not intrinsically about physical motion, such a concept of "speed" can then be applied to provide a model for the change of place through time. On the other hand, Leibniz's phoronomy is distinguished from geometry, as he puts it, because "geometry is imaginary but exact, mechanics real but inexact, and physics real and exact" (GM VI 74). The key difference between phoronomy (which he aligns with physics) and mere geometry is that it treats the concrete elements of extended corporeal motion which are beyond the imaginary entities of geometry. It is not our aim here to argue how such a method purports to provide both reality and exactitude, but the idea here is to narrow the scope of investigation to those *a priori* elements of corporeal motion most conforming to geometrical reasoning such as to provide the grounds of further physical explanation.

The key example of Leibniz's phoronomy was one of his first publicly presented works, the *Theoria motus abstracti* sent to the French Royal Academy in 1671. Towards the end of the *Theoria motus abstracti*, Leibniz characterizes the significance of the work as demonstrations of the "phoronomy of elements" (GM VI 79). This he aligns with the work of Galileo, Fabri, Wallis, and Jungius as important forebears. As we have mentioned, this term is an immediate reference to Jungius' "*Phoronomica sive doctrine de motu locali.*" However the term is also used by Leibniz's early mathematics mentor Erhard Weigel and also by contemporary Leonhard Christoph Sturm.[7] Nonetheless, it appears that Leibniz did not make any special appeal to these works but rather sought to associate his first publication with the themes visited by major thinkers. For what pertains to our inquiry, we take Leibniz's intended solution of the Eleatic paradoxes to be the heart of the text. Like other thinkers of motion before him, Leibniz construes the different Eleatic skeptical challenges to be essentially reducible to a paradigmatic one. For Leibniz, it is the fletcher's paradox. If an arrow in motion is moving towards its target, then it is neither in its (current) place nor where it is not (yet). The problem is that if motion is change of place, then motion is never determinately *in* any place.[8] Motion is then indeterminate.

---

[7] See Erhard Weigel (1776, 197) Leonhard Christoph Strum (1706, 95). Thanks to Stefan Hessbrüggen-Walter for help on these references.

[8] Aristotle's famous refutation of the Fletcher's paradox consists in rejecting that either time or magnitude are composed of indivisible moments or parts (Aristotle 2001, 335; 239b5–9). Although Leibniz does not reject Aristotle's view, he does register his dissatisfaction with it. Since continuous motion must bear some kind of relation with discrete parts, it is insufficient to simply revoke the indivisible. Even if indivisibles are revoked, we still do not resolve how a discrete body is to travel continuously through space in time. Hence in the *Pacidius Philalethes* dialogue of 1676, Leibniz charges that Aristotle "pretended not to see" the problem of the composition of continuous motion (A VI 3, 548; LC 172–173). Thanks to Davide Crippa for his help in clarifying this point.

Leibniz's solution in the *Theoria motus abstracti* is guided by two key concepts, borrowed (with modification) from Hobbes and Cavalieri. For the first concept, Leibniz borrows the concept of endeavor (*conatus*) from Hobbes (Hobbes 1999, 87). Endeavor corresponds to a tendency for a body move and is employed to account for the "start" of a continuous motion that will determine the speed of that motion (in future time). Endeavor could be understood, anachronistically, as a proto-vector or the later dynamical concept of *vis mortua*. But we should be careful not to associate it yet with these two notions. It is, however, warranted in some sense to associate this endeavor with inertial motion, the tendency of a body to persevere within the same motion unless acted upon by another body (or another endeavor).[9] For the second concept, Leibniz borrows the concept of indivisibles from Cavalieri. The concept here is a magnitude smaller than any given (finite) magnitude but which is not geometrical minima. This might remind us of Hellenistic atomism where the atom is a physically indivisible entity but which can still be divided in abstract geometrical terms. Leibniz's version of indivisibles comes down to the idea that, since every determinate thing must be in contiguity with at least two other things, it must have parts, one part in contiguity with one body on one side and another part in contiguity with another side (say, on a line) (GM VI 67–68). With this understanding a point is unextended (non-finite) but is nonetheless a magnitude with parts. That is, if we suspend the usual geometric reasoning for the moment, a point can be understood as both unextended (non-finite) and yet, insofar as it has a "right" and left" side, be divided into two parts. Combining his reading of Cavalieri with Hobbes' work on the definition of point, Leibniz uses this to redefine the nature of the point as a magnitude without extension (GM VI 68). In turn, the two concepts combine into one in the *Theoria motus abstracti*. The endeavor is the motive tendency which describes the indivisible unextended magnitude of the start of an extended motion. In a sense, it is neither the motive tendency (*conatus*) nor the indivisible simpliciter that plays the central role in the *Theoria motus abstracti*. Rather, it is the combination of the two, the geometrical role played by the indivisible magnitude represented physically by the endeavor as the "start" of a motion.

Let us then see how this provides a solution to the Eleatic paradox. We should be careful here because the geometrical terms extension, magnitude, and point all receive unfamiliar and irregular definitions in this early Leibnizian text. Something is unextended when it bears no ratio (no proportion) with finite extension. Something has magnitude when it has spatial parts. Since a point, at least according to this view, has spatial parts (left and right with respect to its position on a line) but no extension, it is an unextended magnitude. Now, we can understand the Eleatic paradoxes to be generated out of the composition of the parts of motions into a continuous whole. Since motion is the change of place (or change of occupied space), each of the parts of motion is the static occupation of the body (in motion) in certain points. Leibniz's argument is that since points cannot compose lines, the positions occupied by a body cannot compose a continuous motion. Hence, given Leibniz's redefinition of the geometrical terms for the sake of phoronomy, we see how the

---

[9] See Mercer 2007, 162 and Beeley 1996, 228–260.

problems are resolved. If we consider motions to be determined by their start, the endeavor (*conatus*), we find that it is unresolvable to a static "point" (GM VI 68–69). Rather, it is grounded in an indivisible magnitude. Hence, no alleged rest separates the many parts of motion. Every part of motion is in motion.

What happens then with the problem of the *ad infinitum* analysis of the continuum into smaller and smaller parts? Leibniz, as we say, "bites the bullet" with this question. With the concept of the indivisible, there is an indefinite division of parts of the continuum. However, with Leibniz's redefinition of the terms, he attempted to avert the contradiction to which this apparently leads. That is, since infinite division is indeterminate, the motion itself is indeterminate. Leibniz's response is that, since the motion of the endeavor is unextended but each point is a magnitude determined in its position by contiguous points on either side of it, this infinite division produces a determinate continuous extension.

Before moving on, let us attempt to review this argument along the lines of the fletcher's paradox. Recall that the fletcher's paradox was meant to induce skepticism about continuous motion by arguing that, since motion is change of place (or change of space) while the arrow can be understood as occupying a series of places (or spaces), motion is itself nowhere. Leibniz's argument in the *Theoria motus abstracti* is ideally suited to responding to this version of the Eleatic paradox in a direct way. That is, since motion is change of place, this change can be attributed to the endeavor (*conatus*) concept. This endeavor is a fully determined magnitude because it is an indivisible situated between extended lines. Hence, motion, or the change of position, is indeed determined spatially albeit it is determined through the unextended magnitude of the indivisible.

The implications of Leibniz's phoronomy in the *Theoria motus abstracti* is the irreducibility of continuity into static discrete points. In some sense we can say that Leibniz's strategy against the Eleatic paradoxes is to provide a bizarre intermediary between continuity and discrete contiguity through the form of the indivisible which is neither fully discrete qua simple nor fully continuous qua finite extended line. What this chimera provides is a terminological maneuver to avoid the obstacles of Eleatic skepticism about motion. Now, in the immediately following years, Leibniz would reject not only the mathematics on which this view relies, rejecting unextended magnitudes (or non-finite quantities) but also the mechanics founded merely on endeavors. However, what remains from this approach is a certain intention behind his strategy against the Eleatic paradoxes.

Leibniz's strategy behind the *Theoria motus abstracti* relies, through the concept of endeavor, on foregrounding the ontological primitiveness of motion. Hence rather than taking motion to be something to be accounted for by spatial positions, he pursued an opposite view. That is, through the concept of endeavor, he saw static geometrical positions as the effect of a more fundamental process of motion. By using the concept of endeavor to designate the start of motion, he ensured that the continuum could be composed of these smaller and smaller unextended magnitudes at every smaller scale. The idea here is that just as every line segment has extrema and every continuous part of that line segment also has extrema, so every extended motion has a start and an end and every infinite subdivided part of that motion also has a start and the end. Since every magnitude of endeavor is determined by a "right" and a "left" part at every scale, the continuum of motion is, to use an anachronistic term, Dedekind-complete (a complete ordered field). The endeavor concept is the "glue" that guarantees the determinateness of any arbitrary division in extended motion. In turn, the notion of a minima (smallest unit of a geometrical division) was rejected at the very start of the *Theoria motus abstracti*. Hence although much of the *Theoria motus abstracti* would be eventually rejected, what remains here as a signature of Leibniz's thought in the decades of philosophical work to come would be the idea of the plenitude of activity. Material things are ontologically actual insofar as there is motion or activity and this activity can be traced to arbitrarily infinite smaller or larger scales.

It is important to distinguish the many aspects of the *Theoria motus abstracti* with regard to what is later accepted and discarded. Leibniz's later self-criticisms of the *Theoria motus abstracti* period make it clear that two key concepts are discarded. The first is the concept of unextended magnitude (not just minima, which was already rejected). With the establishment of the methods of the infinitesimal calculus clearly achieved by 1675, the unextended magnitude was rendered obsolete. What replaces it is the concept of the fictional infinitesimal, which shares with the concept of the unextended magnitude the idea of a continuous variable magnitude whose measure is smaller than any given extension. However, the infinitesimal is a fictional magnitude embedded within the methods of the calculus rather than a real component of physical extension.[10] The second is the concept of physical bodies. With the endeavor concept, Leibniz was able to account for the motion that he saw as essential to the ontological reality of bodies. This was the central thrust of his early ideas about the physical world. However, this motion that, due to the endeavor, allowed for bodies to be real lacked the component of a reactive power of bodies. Hence, in the later writings, Leibniz would pinpoint the lack of the account of resistance, the quasi-inertial properties of bodies examined in earlier chapters, as the central error of the *Theoria motus abstracti*.[11] Nonetheless, Leibniz retains the concept of endeavor in a modified and restricted capacity. It will, as we shall examine below, be associated with rectilinear uniform tendency for motion or *vis mortua*. This is to say that the concept of endeavor will no longer play the central role that it did in the metaphysical foundations of motion even if it remains present in his physical theory.

With the theory of unextended magnitudes discarded and the theory of endeavor demoted to an auxiliary function, what remains of phoronomy? In Leibniz's maturation of the dynamics project, phoronomy continues to play a role but, with the theory of endeavor, it becomes merely one component within the systematic articulation of the dynamics. It is clear that any systematic physical theory must have an account of the coherence of continuous motion in geometric terms. However, this aspect of a physical theory becomes secondary when the geometry of motion ceases to be the fundamental problem of physical theory.

---

[10] For the formal terminology of fictional quantities (*quantitates fictitiae*), see A VII 6, 537; Leibniz 2004, 71.

[11] See GM VI 240; AG 123.

Since causality becomes the central problem of Leibniz's physics in the maturation of the dynamics project, it becomes clear that phoronomy cannot provide much of an answer to this form of inquiry. We should be clear that the phoronomic method itself should not be identified with the content of the physical theory developed by Leibniz during the period of the *Theoria motus abstracti*. The method remains important but subordinated to more fundamental concerns in the systematic construction of the dynamics. Nonetheless, from the perspective of causation, Leibniz's phoronomic methodology in the *Theoria motus abstracti* has traditionally (including Leibniz) been understood as insufficient because of its inability to accommodate concepts of power and work. In the same way that Leibniz had often criticized Descartes' conception of *res extensa* qua inert material existence, he places the *Theoria motus abstracti*, in these self-critical remarks in the dynamical writings, in the same camp. Of course, in 1671 Leibniz already conceived the turn to endeavor as an attempt to overcome the perceived failures of Cartesianism. Nonetheless, the problem Leibniz criticized was the problem of the dilemma between the inertness or innate activity of bodies. Namely, the problem with Descartes and his own earlier work was the insufficient account of the reactive capacities of bodies. In other words, the problem of the *Theoria motus abstracti* in Leibniz's later evaluation was not the problem of continuity but that of an incomplete concept of inertia. Hence, we can see the same kind of explanation for the continuity at work in the later dynamics. That is, the ontological primacy for activity provided the grounds for physical extension. This was made, however, without any reliance on the endeavor concept or indivisibles. But how can Leibniz provide such an account?

Before answering this question, we simply note that the phoronomic method was the main approach for explaining the continuity of motion as long as continuity was taken as a phenomenon that required a geometrical explanation. In this method, once a compelling mapping of physical continuity is made to geometrical principles, the task is complete. What occurs in the dynamics is very different. The dynamical project is based on a rejection of the methodological assumptions on which the *Theoria motus abstracti* depended. That is, the dynamics rejects the logical dependence of the physical coherence of continuous motion on its correspondence with geometrical concepts. This is, of course, not to say that the theory of continuity developed in the dynamics is geometrically incoherent. Rather, the dynamics accepts the mapping of continuous physical motion to geometrical continuity but treats this correspondence with the caveat that the mere geometrical understanding of motion leaves motion indeterminate insofar as merely imaginary. The indetermination of extended motion is indeed an important assumption of the doctrine of *vis*. It is this indirect relationship between continuity and the doctrine of *vis* that we shall examine this in the next section.

4.3   The Indetermination of Continuity

The last concentrated effort to produce a phoronomy in the vein of his earlier projects came at the end of Leibniz's Paris period. Without examining the text in any length, I wish merely to point to the significant distance between this later work and his youthful writings. In the dialogue *Pacidius Philalethes*, composed during his departure from Paris in 1676, by way of England and the Netherlands, back to Germany, Leibniz made an attempt to solidify his presentation of a theory of corporeal motion (A VI 3, n.78; LC 127–221). Here, Leibniz still saw his task as addressing Eleatic-style paradoxes for the reality of motion. His solution was a unique blend of his earlier method and quasi-occasionalism. In this text, Leibniz argued that if motion is change of place (spatial displacement), then this change must be the change of discrete contiguous spatial positions. Since full-fledged continuity engenders the Eleatic paradox because motion occurs neither from where the body is moving nor to where the body moves, motion is determined as that which occurs in the transformation between successive contiguous positions of a moving body. This is the unique theory that he names "transcreation" (A VI 3, 567; LC 212–213). It should be noted that Leibniz, having matured mathematically during the Paris period, no longer appeals to unextended magnitudes for his argument. Nonetheless this solution to the problem of continuity does not recur in his subsequent work since the very question of the determinateness of motion fundamentally shifts away from this geometrically-based form of inquiry (phoronomy).[12]

Of course, the geometrical account of motion or phoronomy would continue to be an aspect of the physical theory but as a separate aspect of explaining the geometry of motion. In the mature work leading up to the dynamics, Leibniz would anchor his theory of motion to the causality of *vires*.

To see how this turn to causality away from geometry transformed Leibniz's thinking, let us jump forward to the end of the dynamics project. In a later stage of Leibniz's career, he writes, in the correspondence of 30 June 1704 to De Volder,

> And there is no reality in anything except the reality of unities, and so phenomena can always be divided into lesser phenomena that could appear to other more subtle animals, and the smallest phenomena will never be reached. By contrast, substantial unities are not parts but the foundations of phenomena. (LdV 303)

In this later stage of Leibniz's career, a few years after what I deem to be the end of the dynamics project, we see a completely different strategy for the accounting of the "greater" and "lesser" in phenomena. The unity that grounds phenomena, provided by the ontological unity found in substance, is orthogonal to the problem of composition. Hence, if we look backwards from the late period, the problem of

---

[12] The term "transcreation" (*transcréation*) does appear in §91 (Part I) of the very late (1710) *Essais de théodicée*. However, Leibniz is referring here to the transition of a non-rational, or merely sensible, soul to a rational one. Although the continuity of the degree of rationality in souls is a significant doctrine in Leibniz's philosophy, transcreation in motion here means something quite different from the context *of the Théodicée* (GP VI 153).

continuity is to be addressed not through the accounting of the mereological relation between part and whole but rather the search for a foundation that is more ontologically basic than what is available through the notions of composition (part and whole). In his very last sentence to De Volder in the correspondence of 19 January 1706, Leibniz states that, "In real things there is nothing but discrete quantity, i.e., a multitude resulting from true unities. Continuous quantity, which is not apparent but exact, pertains to ideal things and possibilities since it involves something indefinite and indeterminate, which is not allowed by the actual nature of things" (LdV 339).

We see here that the indetermination implied by continuity is not something that has to be resolved through fiddling with geometrical terminology. Recalling the distinction between geometry, mechanics and physics invoked above, geometry renders continuity imaginary but exact. The problem ceases to be how continuity should be dissected into determinable discrete parts but rather what possible role the irreducibly indeterminate nature of continuity could play in the dynamics when it is excluded from the "actual nature of things."

What is the relationship that bears between abstract geometrical continuity and the continuity of physical motion? This would be a misleading question for the mature Leibniz. As he explains in the *Nouveaux essais sur l'entendement humain* of 1703–1705 (around the same time as the letter to De Volder we have been considering), there is no essential difference between the continuity in geometry and the continuity in physical motion because the continuity of physical motion is itself an abstract relation. "[T]here is no need to postulate two extensions, one abstract (for space) and the other concrete (for body). For the concrete one is as it is only by virtue of the abstract one..." (A VI, 6, 127; GP V 115; Leibniz 1981, 127). This introduces a whole other perspective in considering the nature of the continuity of motion. Indeed, if the continuity of motion can be simply reduced to the imaginary, geometry alone is capable of accounting for the continuity of motion. As such, the meaningfulness of asking why motion is continuous would be extinguished if not for the lingering concept of *vis*.

In treating the phenomenality of motion, Leibniz constantly refers to the crucial distinction between mere relations and the causes of these relations. This distinction separates motion qua relations and causes qua *vis*. Or as Leibniz puts it, "For even though force is something real and absolute, motion belongs among phenomena and relations, and we must seek truth not so much in the phenomena as in their causes" (GM VI 248; AG 131). Hence, as we have already stated, the continuity of motion is due to its relational character qua phenomena. The question is then to understand how such relations are constituted by the action of *vires*.

For the sake of clarity, it is perhaps worth restating how Leibniz places the *indetermination* of motion in relation to the *determination* of *vis*. As an epistemic problem, Leibniz appeals to the Aristotelean standard that the search for "truth" is to grasp the causes of things. We have seen in the last chapter how the methodology of the equivalence of hypotheses is part of this epistemic search for causes. There is, however, also a metaphysical or ontological aspect to this. As Leibniz states in *De ipsa natura*, "So it must be admitted that extension, or the geometric nature of a body, taken alone contains nothing from which action and motion can arise" (GP IV

511; AG 161). Indeed, this has been Leibniz's constant position against the Cartesians. It should be emphasized here, however, that it is action, and by extension *vis*, that makes motion real. Motion understood as mere modification of position is no real change at all. As he states,

> For in the present moment of its motion, not only is a body in a place commensurate to itself, but it also has a conatus or nisus for changing its place, so that the state following from the present one results *per se* from the force of its nature. If things were otherwise, then at the present moment (and furthermore, at any moment whatsoever) a body A in motion would differ not at all from a resting body B… (GM IV 513; AG 163)

Here, Leibniz deploys the indeterminacy of motion qua continuity in order to demonstrate the necessity of a perduring substantial principle that individuates bodies and gives identity to locomotive events.

This substance-phenomenon distinction is based on a certain reduction of motion to phenomenon. This is an important aspect of how Leibniz saw the so-called "mathematization of nature," a reduction of the geometry of motion to geometry itself. Nonetheless, one of the central aims of the dynamics was to produce another layer of "mathematization," an account provided by the conservation of *vis* through the invariant magnitude of mv^2.

Within this idea of the division of labor between an ontological grounding of the constitution of motion and a phenomenon of motion that is inherently continuous, what then happens to the problem of the composition of motion? In the *Dynamica*, Leibniz emphasizes that the move from points to lines and from lines to planes from a geometrical perspective are constitutive but not compositional. Here, as we have already noted, he emphasizes, "I say [that it is] constituted, not composed" [*Constitui dico, non componi*] (GM VI 370). In turn, when motion is given an ontological grounding through *vis*, it is constitution rather than composition that would guide Leibniz's conception of the relation between points along the path of motion and extended motion.

Hence, although we have already touched on why Leibnizian *vires* is orthogonal or otherwise distinct from Newtonian force, it is important to mark the difference between how Leibnizian *vires* constitutes motion and how Newtonian forces compose extended motion. The latter is continuous insofar as continuous operative forces are at work in the path of a body's motion. Leibnizian *vires* are continuous in a very different way. We can say that Leibnizian *vis* is to be strictly understood as a capacity to produce physical effects. However this capacity of produce effects is not punctual (and not operational) but structural since the accomplishment of work is always mediated by time, *in media res*. For example, the full height reached by a pendulum is the total achievement of the effect of a machine qua work, but this *vis* is a constant at every point in the climb of the pendulum. This structural "continuity" can be very simply understood as the invariance of *vis* in its action in space and time. In order to examine this more clearly, we must examine how the action of *vis* maps onto the continuity of motion through Leibniz's distinction of living and dead *vires*.

In order to dissect the difference between the constitution of motion through *vis* and the composition of motion through (Newtonian) force, we need to examine the distinction of living and dead *vires*.

4.4 *Vires Vivae et Mortuae*

Second only to the measure mv^2, the distinction between living and dead *vires* is perhaps the most well-known aspect of Leibniz's dynamics. It is no surprise then that in Leibniz's most public statements it is this distinction between living and dead force that he draws on to lay out his project. Indeed, by showing the difference between living and dead *vires*, Leibniz points to the usefulness and universality of the theory of *vis*. In addition, the aspect of the dynamics most well-known to historians of physics is the theory of *vis viva* and the ensuing debate over the meaning of the term in the eighteenth century. These debates have little to do with the concept of *vis viva* itself and how it functioned within the system of the dynamics. Hence, just as we have deliberately separated the concept of force from that of Leibnizian *vis*, we have deliberately refrained from reducing Leibniz's dynamics to a theory of *vis viva*, choosing to speak of *vires* generally. In essence, the concept of *vis viva* is only meaningful with respect to its counterpart *vis mortua*. The difference here consists in the difference between treating *vis* from the perspective of extended motion (displacement qua change of place) versus the perspective of a solicitation to motion.

It is difficult to trace the origin of the *viva-mortua* distinction in the dynamics because, although the term *vis mortua* was used as early as 1673 in correspondence with Mariotte and then 1675 in *De arcanis motus*, it did not receive systematic treatment until much later in the 1689 *Phoranomous* and other texts of the period. The term is also conspicuously absent from many of the intermediate milestone texts toward the constitution of the dynamics project in 1689, like *De corporum concursu* and the highly public *Brevis demonstratio*. Nonetheless, the obvious forebear of the *viva-mortua* distinction is Galileo's circa 1600 *Della scienza mecanica: e delle vtilita che si traggono dagl'instromenti di quella* (translated by Mersenne in 1634) where he defines the term "*momento*" [moment] as the "impetus to go downward, composed of heaviness, position and of anything else by which this tendency may be caused" (Galilei 1960, 151).[13] More to the point, in the uncompleted "sixth day" of Galileo's *Discorsi e Dimostrazioni*, he distinguishes between the force of a moving body and that of a "dead weight" (*peso morto*) (Galilei 1898, 334–335, 1974, 294–295). Certainly, the phenomenon identified by Galileo was the same as that which was identified by Leibniz even where "*peso morto*" has become "*vis mortua*." However, without the more systematic dynamical articulation, it is difficult to ascribe this distinction with any rigor to Leibniz before the mature development of the dynamics.

In emphasizing the later arrival of the distinction between *vis viva* and *vis mortua* in the texts constituting Leibniz's dynamics, we wish to highlight its supplementary rather than central role here. In the *Specimen dynamicum* and other texts in the 1690s, we see that Leibniz describes the error of the Cartesians, which he was perfectly able to identify in the 1680s, without the concept of *vis mortua*, as the unrigorous usage of the distinction of the tendency to move and extended motion. Indeed, in the supplement (*Beilage*) of the *Brevis demonstratio*, dated to sometime around 1697 and also collected as "De potentia ab effectu non a tempore aestimanda" in the

---

[13] See also Galileo 1960, 155–157.

Akademie Edition, Leibniz introduces the *viva-mortua* distinction to explain the Cartesian error (A VI 5, n. 3500; GM VI 119–123). In this new formulation of the refutation, Leibniz argues that the Cartesians had simply transposed, erroneously, the proportion of living forces onto the proportion of dead forces.[14] That is, given the statical theorem of the law of the lever, Cartesians and other contemporaries, such as Wallis, simply replaced the dead force of weight with the product of mass and speed. This approach will fulfill what he already called in 1686 the "abuse of mechanics" and what he would later call the "abuse of the doctrine of statics" (GM VI 117; A VI 4, 2027. GM VI 218). In these texts, Leibniz warned caution in the attempt to understand the relation between dead and living *vires*. This caution however seems to have also eluded modern interpreters. I will attempt to rectify this problem here.

Before pointing to the problem among interpreters, we should briefly look at how this *vis viva-mortua* distinction works in the dynamics. As we have seen in previous chapters, the analogy of the proportion of motion in physical systems to statics, and in particular, the law of the lever, was unavoidable. Leibniz and his near contemporaries all relied on an analogy to the law of the lever and the related properties of the center of gravity to make their arguments concerning the laws of motion. We recall here what we have examined in the second chapter: the law of statics is transposed onto the law of motion by taking the product of masses and velocities in pre- and post-collision (rectilinear collisions) and finding the symmetry that conserves this proportion.

$$m_A \left(v_A - v_A' \right)\left(v_A + v_A' \right) = -m_B \left(v_B - v_B' \right)\left(v_B + v_B' \right)$$

This symmetry is what conserves the quantity of momentum (or the quantity of motion with direction) in collision.

We have also examined how Leibniz argued for the limited conservation of momentum without affirming its universality. As we saw, the insufficiency of inertial motion to account for the measure of *vires* was due to the need to incorporate cases like freefall and other "violent motions" which were not derivable (though nonetheless conserved) through the more basic cases. This was the initial reason why Leibniz argued for the superiority of mv^2 as the conserved quantity in the universe. The distinction between *viva* and *mortua* in *vires* map directly onto this distinction. The reliance on inertial motion for the symmetry of the conservation of momentum explains why the law of the lever could be so easily applied to the case of rectilinear collision. In the same way, the insufficiency of inertial motion to derive the measure mv^2 also explains why the distinction between *vis viva* and *vis mortua* is necessary.

In order to proceed, we should emphasize that the measure of *vis* is invariant and is hence applicable as invariant in all physical phenomena, even in cases of "dead weight." Even if the invariant mv^2 holds universally, the problem is that it is difficult to arrive at a measure of *vis* qua mv^2 in those cases. Understanding the universality of *vis* qua mv^2 and the limited conservation of mv does not mean that *vis viva* and *vis mortua* are ontologically distinct. It should thus be emphasized that the distinction of *vis viva* and *vis mortua* must be treated as different aspects of the *analysis* of

---

[14] See also GM VI 239; AG 122.

motion. *Vis viva* describes the action of *vis* in extended motion while *vis mortua* only expresses momentaneous solicitation to move. In other words, the concept of *vis mortua* allows us to apply our understanding of gravitational solicitation as the "starting point" of an extended motion. Recalling the earlier ideas of 1671, just as a point constitutes the extremum of a line segment, *vis mortua* constitutes the "impetus" of a motion that has not yet begun to move. In turn, any motion can be analyzed into smaller line segments at every scale. In turn each of those line segments, smaller than can be assigned, will have (at least one) extremum. In this sense, the distinction between *vis viva* and *mortua* allows for the analysis of the continuity of motion without the reciprocal idea of the composition of motion from tendencies to move (tendencies which are themselves immobile).

Keeping the aim of the argument in mind, let us look more precisely at how the distinction is presented. We take our statement of the distinction between *vis viva* and *vis mortua* from one of its clearest exposition, the 1695 *Specimen dynamicum* (Fig. 4.1).[15]

> Consider tube AC rotating around the immobile center C on the horizontal plane of this page with a certain uniform speed, and consider ball B in the interior of the tube, just freed from a rope or some other hindrance, and beginning to move by virtue of centrifugal force. It is obvious that, in the beginning, the *conatus* for receding from the center, namely, that by virtue of which the ball B in the tube tends toward the end of tube, A, is infinitely small in comparison with the impetus which it already has from rotation, that is, it is infinitely small in comparison with the impetus by virtue of which the ball B, together with the tube itself, tends to go from place D to (D), while maintaining the same distance from the center. But if the centrifugal impression deriving from the rotation were continued for some time, then by virtue of that very circumstance, a certain complete centrifugal impetus (D)(B), comparable to the rotational impetus D(D), must arise in the ball. From this it is obvious that the nisus is two-fold, that is, elementary and infinitely small, which I also call solicitation, and that which is formed from the continuous or repetition of elementary nisus, that is, impetus itself. (GM VI 238; AG 121)

Not unlike a suspended weight, centrifugal-centripetal force on a tethered rotating ball presents a case where a tendency to move is restricted from actual motion. Now, in this case the rotational acceleration is bracketed for the simple consideration of the ball along the length of the tube. Certainly here, centrifugal-centripetal force is kinematically expressed by the tendency of the ball to move along the tube. This tendency, a *nisus*, or impetus, in Leibniz's terminology, is proportional to the rate of rotation. The distinction between the actual rotation and the impulse (*nisus*) of the ball to move along the tube is then treated through the difference between *vis mortua* and *vis viva*. On this, Leibniz explains:

> One force is elementary, which I call dead force, since motion [*motus*] does not yet exist in it, but only a solicitation to motion [*motus*] as with a ball in the tube, or a stone in a sling while it is still held in by the rope. The other force is ordinary force, joined with actual motion, which I call living force. (GM VI 238; AG 121)

---

[15] Our argument here is developed from the 1695 *Specimen dynamicum,* but the exposition of the distinction between dead and living *vires* is already fully developed in the 1689 *Phoranomus seu potentia et legibus naturae*. The key distinction is made in §18 of this text (Leibniz 1991, 477–478, 2007b, 742–743).

Fig. 4.1 Leibniz's illustration of dead and living *vires* using rotating tubes

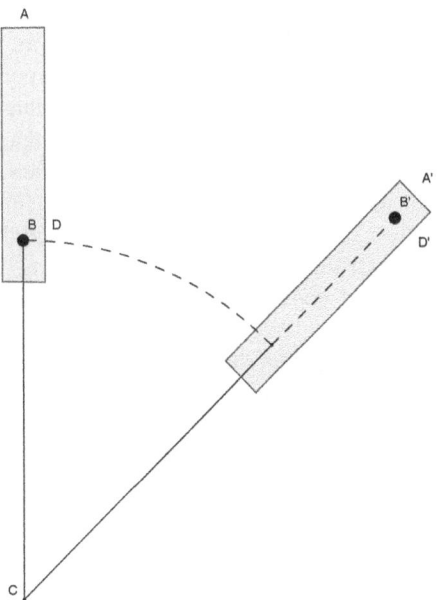

The simplest way to distinguish between *vis viva* and *mortua* is to distinguish between a mere solicitation to move (as in centrifugal-centripetal force or gravitational attraction) and the aggregate effects of these solicitations in actual motion. It is hence easy to understand *vis viva* as the composition of *vires mortuae*. Leibniz provides ample evidence to understand the distinction in precisely this way when in the following passage he writes that,

> [W]hen we are dealing with impact, which arises from a heavy body which has already been falling for some time...the force in question is living force, which arises from an infinity of continual impressions of dead force. And this is what Galileo meant when he said, speaking enigmatically, that the force of impact is infinite in comparison with the simple nisus of heaviness. (GM VI 238; AG 121–122)

This passage seems to suggest that dead force, through "an infinity of continual impressions" gives rise to living force. In the discussion of this in the *Phoranomus*, as in a few other places, he directly invokes the analogy of the "continuous impression of new conatus like the acceleration of gravity" to the geometrical relation of the point to the line (Leibniz 1991, 478, 2007b, 744–745). Reflecting now on gravitation and centrifugal-centripetal force, we might be led to believe that *vis viva* is composed out of the infinite number of *vires mortuae*, just as freefall is composed of the continuous solicitation of gravity, which kinematically describes the acceleration of freefall.

Things are not as they seem. We have noted the use of dead force (*vis mortua*) qua dead weight (*peso morto*) in Leibniz's earlier writings. This was drawn from Galileo. In his later writings, which we are now considering, the "enigma" that surrounded this concept appeared in a different context. This "enigma" is now placed within the context of a mature understanding of the labyrinth of the continuum. Recall that Leibniz held to a theory of indivisibles in his early work, *Theoria motus abstracti*, where the "start of motion," understood through the quasi-Hobbesian endeavor, was understood as a real physical indivisible. In this way, the endeavor at the "start of motion" and extended motion were proportional as a point to a line. The labyrinth of the continuum is just, in a sense, the simultaneous maintenance of this view and the paradoxical (or paralogical) composition of a line through points. To undo the labyrinth, indivisibles, real infinitesimals, and other things of the sort must be given up as real. Leibniz's solution, developed through his work on the infinitesimal calculus, was to give infinitesimals a "fictional" status. That is, an infinitesimal variable quantity or magnitude is simply a quantity smaller than any given.[16] This definition, which he calls "syncategorematic," does not describe a determinate entity or an assignable magnitude but plays an operational role in the *methodology* of the infinitesimal calculus. Hence, though we can use points for the infinitesimal analysis of a curve, they cannot in any way "constitute" a line. The strict implication here is that there is no determinate or assignable ratio between a line and the point. This is a fairly conclusive resolution to the labyrinth of the continuum achieved by Leibniz during his Paris period, and it is through this that the supposed ratio between *vis viva* and *vis mortua* must be understood.

Understood through Leibniz's resolution of the labyrinth of the continuum, Galileo's supposed "enigma" now appears clear. The enigma is not that the "infinity" of the force of impact should have a ratio with the finitude of "simple nisus." Rather, the enigma is that there is no possible (or no assignable) ratio between infinity and finite.[17] This means, in turn, that the continuous solicitations of gravitational attraction of a body in freefall are simply that, continuous, and hence bear no assignable ratio with the simple "nisus" of weight. Hence weight and freefall can be analogized through the relation between a point and a line or the finite and infinite (or infinitesimal and finite), but only if we understand that there is no assignable ratio in the two cases.

These passages in the *Specimen dynamicum* are indeed confusing, but we can take a second approach to the analysis of this distinction of *vis viva* and *vis mortua*. Here we examine the famous problem of why Leibniz did not include the factor of 1/2 in his estimation of *vis viva* as mv^2. The standard response is that the measurement of mv^2, as we saw in Chapter 2, comes directly out of the proportion $\Delta v^2 \propto \Delta h$. With this, what is at stake is the proportion between velocity and work rather than the "absolute" measure of energy-work itself. This is accurate, but it indicates a further conclusion. It indicates that the quantity mv^2 is not the result from the integration of "continual impressions" of *vires mortuae* as it might have initially seemed. If this was not the case, the coefficient 1/2 in $1/2mv^2$ would have been a natural conclusion.

---

[16]This methodology of using infinitesimals smaller than a given variable magnitude, a given error, is definitively developed and employed by Leibniz at least by 1675 in *De quadratura arithmetica circuli ellipseos et hyperbolae cujus corollarium est. trigonometria sine tabulis*, and Leibniz will continually return to this concept in his later presentation of the infinitesimal calculus and other letters defending and justifying its method (Leibniz 2004). See the often quoted letter from Leibniz to Pinsson of September 1701 (A I 20, 492–494).

[17]In some texts Leibniz also uses the expression "larger than any finite number assignable" (A VII 4, n.12; n.16). This "more" absolute notion conflicts somewhat with the "variable" notion of the infinite/infinitesimal here as a quantity smaller than a given assigned quantity. However the different contexts suggests that the "variable" notion becomes prioritized in central methodological texts like *De quadratura arithmetica circuli ellipseos et hyperbolae cujus corollarium est. trigonometria sine tabulis*. See Knobloch 2008, 176.

Since the velocity in time is Δs and acceleration in time is Δv:

$$\Delta E \approx m \cdot v \cdot \Delta v$$

On the other hand, if we *integrate* over the changes of v then:

$$E = \int_{s0}^{s1} m \frac{dv}{ds} \frac{ds}{dt} ds = \int_{v0}^{v1} mv dv = 1/2mv_{v0}^{2v1}$$

Or simply:

$$\Delta E = \tfrac{1}{2}m\Delta v^2 = \Delta\left(\tfrac{1}{2}mv^2\right)$$

Now, if this form of integrating over the factors of dv seems to privilege a Newtonian understanding of momentaneous forces, we could also see the same proportion arise from a linear equation using only maximal values.

Evaluating ΔE through $m \cdot v \Delta v$ means treating $v \Delta v$ through average velocity:

$$\Delta s / \Delta t = \left(v_{intial} + v_{final}\right)/2$$

$$F\Delta s = m\left(v_{final} - v_{intial}\right) \cdot \left(v_{intial} + v_{final}\right)/2$$

With initial velocity = 0 and final velocity = v:

$$F\Delta s = mv^2/2$$

Commentators have argued that Leibniz was in the habit of dropping constants.[18] While this might be true, coefficients play a role in algebraic structure and derivation that can hardly be compared. The idea that the "1/2" is simply dropped does not get at the heart of the problem and, at least for our purposes here, does not illuminate the relation between *vis viva* and *vis mortua*. In fact, my argument here is that the "1/2" was not "dropped" but simply did not arise since, whether one takes the route of integration or linear equations, the coefficient 1/2 is nearly unavoidable.

This unavoidability of the proportion of "1/2" is part of the evidence we are indicating here for the idea that there could be no proportion between *vis mortua* and *vis viva*. In Bertoloni Meli's classic work on Leibniz and Newton, he convincingly provides a solution for how we should understand the analogy between *vis viva* and *mortua*, and line and point. The solution is roughly that the infinitesimal-finite (or finite-infinite) difference provides the analogical terms for comparing the difference in dimensionality between *vis viva* and *vis mortua* without implying that the former is an integration of the other. It seems, however, that this integration is exactly what is occurring when Leibniz explains that,

---

[18] Cf *Duchesneau* 1994, 226–231 and Gueroult 1934, 38–48.

[W]hen we are dealing with impact, which arises from a heavy body which has already been falling for some time, or from a bow that has already been restoring its shape for some time, or from a similar cause, the force in question is living force which arises from an infinity of continual impressions of dead force. (GM VI 238; AG 122)[19]

Indeed, the ambiguous phrase "arises from" indicates that the continual or continuous impressions of dead force are constitutive of living force. What we wish to underline here is that this constitution does not rely on the method of integration and does not imply a ratio between the two. Here Bertoloni Meli argues that, "When he talks of a 'heavy body which has been falling for some time', he does not mean that the integral of dead force is multiplied by some element of time, but is simply providing a general description of the phenomenon" (Bertoloni Meli 1993, 90). This understanding of the analogy between *vis mortua* and *viva* and infinitesimal-finite magnitudes as a "general description" of phenomenon allows us to reconsider the mathematics of continuity at work in the dynamics. That is, through a process of differentiation, we are able to determine moments of motion (momentaneous motion) from extended motion, but the recomposition of motion from momentaneous motion does not follow. The role of these different kinds of infinitesimal and finite terms is thus only to compare these quantities of different dimensions. Hence, we find Leibniz, in a 27 December 1698 letter to De Volder, arguing that,

Of course the speed increases in equal amounts according to time, but the absolute force itself increases according to distance or the square of the times, i.e., in accordance to the effect. So by analogy with geometry, or my analysis, solicitations are as dx, speeds are as x, and forces [*vires*] are as xx or $\int$xdx. (GP II 156; LdV 33)

Here again, what is misleading is the idea that solicitations integrate into *vis viva* qua $\int$xdx. What is at work, however, is only the use of these quantities to analogize the dimensionality of the magnitudes involved in the measure of speed and *vis*.

Leibniz's dynamics thus does not include a theory of the positive geometrical proportion between *vis viva* and *vis mortua*. As we have seen, the many instances of his use of the finite-infinitesimal relation to analogize the dimensionality of *vis viva* and *vis mortua* would have provided ample opportunity for such a theory to take shape. This indicates a methodological limit in Leibniz's dynamics rather than an error in calculation or the dropping of constants. That is, it indicates that the concept of *vis* has no direct kinematic counterpart through momentaneous motions. The continuity of motion is thus not built up from the integration of infinite *vires mortuae*. To the contrary, it is the continuity of motion which allows us to develop an analysis, via the infinitesimal calculus, to determine the momentaneous effects of *vis* distributed in continuous space and time. We shall examine this claim further in the next section in order to provide a deeper appreciation of the role of continuity in the dynamics.

---

[19] In a different context, Leibniz also argues for the same distinction between dead and living force to De Volder in a letter of 27 December 1698. Here, Leibniz straightforwardly claims that the analogy to the distinction between finite and infinite for the distinction between dead and living force is made to argue for the continuity between the terms on the model that *natura non facit saltum* (GP II 154, LdV 33).

4.5   The *Ex Hypothesi* **Continuity of Motion**

In the last section, we have argued that *vires mortuae do* not integrate into *vis viva* despite all appearances to the contrary. Of course, just as a line can be analyzed through its points, extended motion can be analyzed through its moments. This does not mean that points can compose a line any more than *vires mortuae* can compose *vis viva*. Part of the problem was Leibniz's rejection of a determinable or assignable ratio between infinitesimal and finite and, with this, the rejection of any actual infinitesimal entity. Now, neither the methodology of differentiation nor integration commits us to an actual infinitesimal or assignable ratios between infinitesimal and finite. However, the failure of establishing a relation of integration between *vis mortua* and *vis viva* indicates an important aspect for our understanding of the concept of *vis* and its relation to extended motion. That is, there is no direct kinematic counterpart of *vis* and as such all quantitative evaluations are necessarily phenomenal and hence essentially and conceptually distinct from *vis* itself.

What would it mean for *vis* to have a kinematic counterpart? By this I mean to ask how *vis* can be expressed as a component of the "kinematics" or geometrical analysis of motion. With this we make more precise what it means for *vis* to produce continuous motion. That is, we ask the question: how do parts of motions reflect the action of *vis*? It is commonplace to repeat the platitude that victors write history but with respect to this question in Leibniz's dynamics, the "victory" of Newtonian mechanics has often obfuscated an appropriate interpretation of this question in Leibniz's dynamics.

The key obstacle produced by the habit of reading Leibnizian dynamics against the backdrop of established Classical-Newtonian mechanics or dynamics is the treatment of forces as the kinematic element of a tendency to move in a proto-vectorial sense. As we have emphasized frequently, velocity or momentum change is simply orthogonal to the conception of Leibnizian *vis*. This also follows from the interpretation just refuted. Is accelerated motion the result of an integrated sum of dead forces operating continuously in space and time? This would render Leibnizian *vis viva* into a kind of Newtonian force. In this view, since the action of gravity on a suspended body constitutes the "dead" weight of the body and the acceleration of a body in freefall constitutes the "living force" of a body, the concept that would most conveniently unite the two cases is Newtonian force. Gravitational force, understood in the Classical-Newtonian manner, impresses onto falling bodies the change in momentum that allows for the acceleration that results in Galileo's law of falling bodies. As Newton puts it in his second law of motion, "The change of motion is proportional to the motive force impressed; and is made in the direction of the right line which that force is impressed" (Newton 1972, 13). In this view, the central epistemic role played by force is to account for the continuous activity of forces, understood roughly in contemporary terms (due to Euler's formulation) as $\vec{F} = m\vec{a}$, that result in the kinematic path that a body takes in motion. After our analysis of the relation between *vis viva* and *vis mortua*, we can see that this view is entirely absent in Leibniz.

The lack of a kinematic account of motion might push us to say that Leibniz's dynamics was no dynamics at all. Insofar as Newtonian force qua cause of momentum change is to be sharply distinguished from Leibnizian *vis* qua cause of the accomplishment of work, we can say that Leibnizian dynamics aimed at an entirely different problem than the Newtonian one. In the last chapter, we saw Leibniz's reduction of all rotational motion to rectilinear ones in order to defend his conception against Newton's challenge of rotational motion. We underlined, however, that Leibniz's strategy concerned the reduction of one form of motion (curvilinear) into another form of motion (rectilinear) rather than one form of force (dead) to another (living). Despite the flaws of this view, the fallacy of composition, the conception of *vis* was itself never at stake. The lack of a kinematical account, or rather, the lack of a kinematical role played by *vis* should indicate a methodological gap in Leibniz's dynamics rather than an error, implicit omission, or some such failure.

The positive methodological picture provided by Leibnizian dynamics is one that begins with *vis*, the structural property of a system of bodies, a principle of conservation rather than a principle of modification. In turn, through the theory of *actio*, *vis* is distributed in space and time in terms of velocities at certain times and the mutual positions between bodies at certain times. As such, the dynamics is based on a methodology of statics, tied to the proportions of extremal quantities such as maximum height, maximum work, and maximum velocities. We have seen how Leibniz attempted to move beyond statics by providing an account of the activity of *vis* in time (the concept of *actio*). However even in those accounts, like in the *Essay de dynamique* discussed in Chapter 2, the goal was to demonstrate the conservation principle still essentially based on an expanded conception of the center of mass frame and still reliant on the static proportions between work and motion. Leibniz's theoretical ambitions thus outstripped the methodology available to him. It is through this that a lacuna is evident. Hence, the comparison of Leibniz's method to Newton's reveals an error of interpretation, but the consideration of Leibniz's method to his conception of *vires* reveals a limitation in this methodology.

These limitations of the dynamics are, again, due to its continued reliance on statical methods. Indeed, if parts of motion are to be analytically treated through *vis viva*, we see that momentaneous parts of motion behave statically. A momentaneous impetus or endeavor, mv, remains analytically identical to the simple tendency to move in a suspended body or centrifugal force. The sum of static moments is thus unable to represent the transformation of these moments in the path of motion. Again, just as points are useful for the analysis of curves, they cannot compose curves. In this sense, *vis mortua* cannot stand in as an analogue for Newtonian force and its composition.

Hence, the result of this analysis is that the measure of *vis viva* arises in motion only through the continuous extended motion rather than from any composition from non-continuous parts or moments. It is one thing to say that motion is constituted by momentaneous tendencies to move, but it is quite another to say that these tendencies compose this extended motion. In this sense, the dynamics has no account of the composition of the path of motion. Rather, the key claim of the dynamics, the conservation of mv^2, requires that motion be more than discrete moments of *vis mortua*, it assumes continuous motion and incorporates the transformation of moments of motion (acceleration) in the achievement of work.

The indirect relation between the theory of *vis* and the phenomenon of continuous motion thus establishes that the continuity of motion in the dynamics is a structural feature of motion rather than due to any operational property. We might also say that the continuity of motion is generic or transcendental. In this case, we say that continuity is a generic property of motion because it is a property of motion per se. In turn, continuity is a property of motion insofar as motion is admissible to geometrical analysis. What thus accounts for the continuity of motion is its mathematical analysis rather than its ontological composition. The issue here is not that that a composition of continuous motion is not available in a strict sense. The problem is that a composition of continuous motion is available only through a composition of other smaller parts of continuous motion. Rather, the composition of continuous motion is not available from the elements of non-continuous parts or moments. We saw that the early work of Leibniz did contain attempts to account for continuous motion from non-continuous parts. This was seen as necessary because of the need to resolve the indetermination introduced by continuity into non-continuous parts. However, once Leibniz shifted to an argument for the determination of motion on the basis of a constitutional rather than a compositional foundation, continuity could be treated as the determinate property of phenomena responsible for the variability of relative motion.

The argument in this chapter comes down to the question of whether abstract geometrical continuity can be separated from concrete physical continuity. Since Leibniz in his mature period saw only metaphysically actual substance as discrete, it appears that all continuity is an abstraction. As he often repeats in his mature writings, "[T]ime, extension, motion, and the continuum in general, as we understand them in mathematics, are only ideal things–that is, they express possibilities, just as do numbers" (GP IV 568, L 583). The ideality of mathematical continuities is separated from the actuality of metaphysically real unities.[20] However, in the maturation of Leibniz's physical theory, this abstract or ideal nature of the continuity of motion contributes in a fundamental aspect to the dynamics. It provides a phenomenal form in which *vis* is expressed. The question of whether motion is actually compositionally discrete or continuous is thus wrongly posed. Motion, qua phenomenon, is necessarily continuous. Yet this "necessary" continuity is conditioned on the premise that the theory of *vis* provides a more foundational layer of structural explanation. As such, we can establish an ontological hierarchy between cause and effect in the dynamics. In this analysis then, it becomes explicit that cause and effect belong to different ontological levels. Causes, *vires*, are genuine unities generating the phenomena of motion which is, in turn, composed of continuities. What is caused, the effect of extended motion, is a collection of continuous parts insofar as no sequence of division can provide a final minimum part.

---

[20] Recall Leibniz's famous "ontological maxim" stated to Arnauld years earlier: "[W]hat is not truly *one* being is not truly a *being* either" (GP II 97).

4.6 Concluding Remarks

In the conclusion of this chapter, I wish to briefly explore a speculative dimension of the problem left behind by the interpretive work here. We have examined the role of continuity in Leibniz's dynamics through how he shifts from searching for a principle of determination for continuous motion from geometry to the role that *vis* plays in allowing continuous motion to be grounded in something determinate. The role continuity plays here allows us to establish the heterogeneity of cause and effect in order to solidify our claim that the level of causation, *vires*, and the level of effect, extended motion, are ontologically distinct but causally related.

In the above I have treated *vis viva* as a whole governing parts of continuous motion insofar as it is an individuating condition for a locomotive event. The reason for this is that the phenomenon of motion is subject to the relationalism of motion understood through the equivalence of hypotheses. It is the ontological foundation of *vis* and this theory of variation that grants this motion its determination through an underlying ontological foundation. In this view, the account of motion through *vis* does not treat motion as a literal totality. That is, although *vis* individuates a locomotive event, it is not the total sum of its parts. These terms are not homogenous. As such, the *measure* of *vis* is relative to a physical system as an *abstraction*. Every physical system is embedded in a larger physical system just as each physical system has embedded within it a smaller one. The exception here is the created universe taken as a whole.

Now, before thinking about the physical universe as a whole, we notice that just as the measure of *vis* holds for a medium-sized physical system, it holds for the physical system as a whole. The conservation of *vis* is universal and scale-invariant. This is why the measure of *vis* for a physical system does not have to take into account the infinite number of different sub-motions within the consideration of a physical system of a handful of medium-sized bodies. In the dynamics, the difference here is given the name "relative or proper" *vis*, for the internal motion of a body, and "directive or common" *vis*, for the relation between bodies that constitute a physical system (GM VI 238–239; AG 122). Hence, if we take the totality of the created universe as a whole, there is no "common" *vis*, only "proper" *vis*.

Would the motion of the universe as a whole be continuous? This question is, in a sense, wrongly put because motion can only be taken as relative change of place. The universe as a totality would not move at all. This reason for this is simple. The universe as a whole does not stand in a spatial relation with any other thing. From this it follows logically that since it does not move, there would be no way to measure its *vis*. Putting this in other words, when we consider the universe as a whole, there is no distribution of *vis* among its parts of motion. The moment we consider the "modification" of *vis* in its distribution among various bodies in a physical system, we step back from the totality of the world. *Vis*, from the perspective of the totality of the world, is unmodified.

The fascinating aspect of this is that, because *vis* is not subject to modification, the totality of the world is a perpetual motion machine.[21] Recall here that the rejection of perpetual motion, Torricelli's principle, was a necessary condition for the measure of *vis*. Here, if we are considering the totality of the universe, conservation of *vis* means that the total magnitude of *vis* in the world can be arbitrarily increased or decreased without modifying the motion of the universe as a whole. The perpetual motion of totality can be seen as a trivial result from the fact that the totality is actually neither moving nor at rest since motion can only be a relative property between two or more externally related bodies. However, it also suggests, less trivially, that phenomena can only be produced with respect to non-total systems of the world. A totality is essentially static and is hence incapable of expressing action, the translation of *vis* into motion (in space and time).

From this speculative perspective, we can understand the continuity of motion as *necessitate ex hypothesi* for an important reason. Recall that Leibniz aligns discreteness with actual things and continuity with imaginary ones. Now it appears that the domain of physical phenomena arises only with respect to non-total systems of the world. Hence, we can say that the phenomenon of motion arises only when the physical system in question is non-total, the property of motion is categorically irrelevant when the totality of the physical world is considered. Motion can thus only be a property of non-total physical systems where Torricelli's principle holds. As such, the principles of motion and the principles of *vis* really do constitute different ontological domains. Although *vis* is capable of determining the principles and structure of motion, it is not homogenous with the phenomenon of motion itself.

---

[21] See Stan (2016b).

Chapter 5
The Changing Concept of the Equipollence of Cause and Effect

Abstract This chapter concludes the three-part presentation of the central architectonic components of the dynamics. In this chapter, we examine the most important (and most explicit) principle of the dynamics. The equipollence of cause and effect, inherited from Scholastic thinkers but creatively reinvented by Leibniz, can be understood as the starting place of the dynamics project. In this chapter, we trace Leibniz's different interpretations of this principle in order to move from a dynamics based on final causation (teleology) to that of formal cause (*via* the concept of *actio*). It is through this development that the very concept of *vis* matures in Leibniz. By examining this development, this chapter provides a fuller understanding of structural causation in the dynamics.

5.1 Introduction

In the last chapter, we ended on a speculative note. We considered how the concepts of the dynamics might be applied to the universe taken as a whole. There, we touched on a tension that has been running throughout the book. The central argument that we have been developing is that of structural causation. That is, the causation involved in the dynamics is a cause that bridges different ontological realities, that bridges a dynamical level of causes and a phenomenal level of extended motive effects. The tension we touched on was whether we could consider causation on the level of extended physical reality only. We considered this possibility from the perspective of the totality of the physical universe (the totality of extended physical reality). We left off with the speculation that, since continuity only arises in the consideration of the relation between parts of a whole, it is irrelevant to the consideration of the whole as a totality. A totality can be said to be neither in motion nor at rest, a totality cannot be phenomenal. Coherent physical phenomena depend on the mutual relation between bodies and a totality has no external relation against which these properties can emerge. We can leave this speculation behind as a kind of limit case for the concepts developed in the dynamics. Nonetheless, it does point to the following tension within our interpretation. Could there be a dynamics based merely on phenomena (size, shape and motion)? If dynamics purports to account for motion from the perspective of causes, could we have a theory of causes based on purely mechanical processes? If *vis* is simply the totality of "energy" in the physical universe as a whole, would we simply be able to frame causation as purely mechanical and do away with the more complex proposal for a theory of "structural causation"?

This alternative explanation of the dynamics is a very plausible one. Indeed, Leibniz for many years treated *vis* almost exclusively as *potentia*, an intensity of energy that is exhausted in its accomplishment of work (through motion). Now, if a smaller physical system is embedded in a larger one, the smaller system will surely have strictly "less" *potentia* than a larger one. *Potentia* would still be a conserved quantity for any physical system taken as closed and hence a structural property. That is, the measure of *potentia* can only be measured by examining a physical system as a whole through its entire range of motion.

The alternative view also allows us to provide a sensible explanation for the appeal to the "incorporeal" or "metaphysical" within the dynamics. Even in Leibniz's mature presentation of the dynamics, the "metaphysical" and "mechanical" are often neatly divided into the domain of final causes and efficient causes. The "metaphysics" of the dynamics concerns conservation itself, a final cause determining the choice of a conservation quantity. In many of these presentations, Leibniz remarks that the principles of the dynamics could only be grasped by metaphysical meditations rather than empirical observations. However, the point is that the metaphysical aspects would concern only the determination of the laws of corporeal motion and collision. While the notion that God chooses among different physical laws in his creation of the actual world is a salient aspect of Leibniz's maturing philosophical perspective, interpreting the "metaphysical" aspect of the dynamics solely from this perspective reduces the principle of conservation to that of an abstract and mediated final cause.

This view is coherent and plausible. It neatly divides the dynamics into what Leibniz calls the kingdoms of power and grace. As he writes in the *Specimen dynamicum*,

> In general, we must hold that everything in the world can be explained in two ways: through the *kingdom of power*, that is, through *efficient causes*, and through the *kingdom of wisdom*, that is, through *final causes*, through God, governing bodies for his glory, like an architect [...] These two kingdoms everywhere interpenetrate each other without confusing or disturbing their laws, so that the greatest obtains in the kingdom of power at the same time as the best in the kingdom of wisdom. (GM VI 243; AG 126–127)

The distinctness of the kingdom of power and that of wisdom certainly seems to prescribe a neat separation between mechanical relationships between bodies and the metaphysical determination of laws.

My concern here is not to directly refute this picture but to argue that, alongside this view of interpenetrating kingdoms, there is another causal theory that is more fundamental to and paradigmatic of the dynamics project as a whole – but also more difficult to defend. As I have already argued in a previous chapter, I take structural causation to be a theoretical extension of the structural property of conservation. As final cause, conservation is grasped virtually, that is, through the completion of the "future effect" of motion. However, in the establishment of the dynamics properly speaking in 1689, Leibniz embraces a theory of *actio* that replaces the centrality of *potentia*. The concept of *potentia* is reducible to the theory of *vis* as final causality. *Potentia* is a physical causal property that produces effects through its exhaustion through work. It is thus inseparable from the notion of "future effect." This virtuality or future-oriented notion of *potentia* was certainly crucial in Hobbes (Hobbes 1999, 101). The concept of *actio*, I argue, is an alternative concept precisely because cause and effect are temporally actual at every moment of a physical system's achievement of work. We have already given a sketch of why this is the case in the second chapter. Our task here is to provide a more direct account of what this concept of *actio* means for understanding causality in the dynamics through a critical examination of the equipollence of cause and effect.

In Leibniz's many presentations of the dynamics after the initial coining of the term and the composition of its main treatise, the *Dynamica*, we find this division between final and efficient causes that appears to neatly separate the dynamics between a metaphysics that concerns only the determination of laws (from the divine architect) and a mechanics that accounts for the operation of bodies. However, at the heart of the dynamics is the problem of the action of bodies (or bodies as a physical system). As such, dynamical cause is concerned with how *vis* is translated into motion not only as a mechanical process which reflects a conservation law. Rather, it is based on the inherence of *vis* within physical systems. As such, since both efficient cause and final cause are reflected in external features of motion, the former is a transformation of the measure of extensional magnitudes (motion and mass), the latter is an abstract conservation of a system, only *actio* can provide an account of causation with respect to the inherent and actual behavior of a body or physical system in the accomplishment of work. *Actio* is then readily recognized as a formal cause of corporeal motion. From this notion of formal causation, the concept of effect is changed from that of "future effect" to the actual and continuous effect, which is locomotive phenomenon. The generation of the phenomenon of motion, within the dynamics, is a factor of both velocity, measured relatively between bodies within a physical system, and accomplished work, also measured within the domain of the physical system in question. The problem of intensity still remains, of course, but it becomes a subordinated feature of the dynamics relevant only to the translation of "energy" to motion within a physical system and thus part of the account of phenomena. It is through *actio* that *vis* becomes an invariant property across different physical scales. Hence, *actio* becomes the formal cause of motion, the constant cause of locomotive phenomena responsible for the distribution of motion across space and time. With the same stroke, *vis* qua *actio* becomes the cause of motion not only in the distant and abstract sense of a teleological choice of the law of motion and also not in the sense of mechanical relations, but an active metaphysical cause that continuously determines the production of motive effect qua phenomena.

Now, having given a sketch of the argument that will follow, we should underline that this development of the concept of *actio* transforms the nature of causality within the dynamics from one where cause and effect could have been understood as operative through homogenous domains to one where causality works across different ontological registers. In this, we should also highlight that this can most readily be seen from the perspective of what it means for extended motion to be an "effect." In fact, this is where our investigation here begins. The key point is that if motive effect is only extended motion, Leibniz must remain tied to the concept of causality as that which brings about future effect. Once effect also includes temporally evolving states of velocity, the relation between cause and effect becomes an immediate relationship governed by the proportion between velocity and achieved work. The result is that *vis* qua cause produces effects continuously that, although harmoniously convergent with the final cause of conservation, is an on-going actualization that does not depend on the virtuality of a future effect. At the same time, *actio* is not a mechanical operation since it does not depend on the extensional properties of motion and interaction. Through the concept of *actio*, Leibniz realizes a theory of formal cause of motion inherent in physical systems that reduce neither to final nor efficient cause. As formal cause, *actio* is hence invariant on any spatial and temporal scale. In this sense, *vis* qua *actio* is structurally causal insofar as its effect determines the actual and continuous production of phenomena of any physical system across spatial and temporal scales.

In this chapter, we will first revisit the initial meaning of the concept of *vis* as *potentia*, based on the superiority of the conservation of work over the conservation of the Cartesian quantity of motion. It is this concept that first answers to the relationship between cause and effect established by equipollence. We shall then examine why Leibniz saw this as an "incorporeal principle" superior to the consideration of motion from purely mechanical considerations. It is this path of thinking about the nature of *vis* qua cause that would lead Leibniz to slowly bring another conception of *vis* to the fore. That is, what *vis* causes in the world is phenomena. Hence what we witness in the maturation of the dynamics project is the transformation of the concept of effect and what it means for *vis* to act within the world. It is in the development of the concept of *actio* that Leibniz would present a mature theory of *vis* as the cause of motion, a concept that would accomplish his lifelong goal to account for corporeal motion through an intrinsic causal principle in bodies. Finally, we shall spell out how this constitutes a theory of structural causation.

5.2 *Vis* qua *Potentia*

As we have remarked in the second chapter, the dynamics project can be traced to the formulation of the principle of equipollence and the measurement of *vis* in the texts of 1676 and 1678. From this point on, Leibniz will constantly refer to the principle of equipollence as an "axiom" for thinking about the theory of motion in general and mechanics specifically. What we have not examined until now, however, is the oddness of the adoption of the principle of cause and effect in the context of these earlier texts. Why should the adoption of the principle of equipollence lead Leibniz to posit a principle of motion other than the conservation of the quantity of motion (mv)? The principle of equipollence states the relation between a capacity to achieve a certain effect and the achievement of that effect in time. It is this notion that Leibniz emphasizes in *De arcanis motus* when he argues, "[I]t is therefore necessary that cause and effect are perfectly resolved into the same thing in the end" (A VIII 2, 60). We have noted in previous chapters that the relation between cause and effect is to be found in the "identity" of motion and work in time. However, this relationship between the principle of equipollence and its implied notion of the capacity for future work does not allow us to decide between the conservation of the quantity of motion (mv) or some other conservation. In what immediately follows, I will give an account of how Leibniz bridges the gap from the theory of cause and effect to a different concept of conservation qualitatively beyond the conservation of the quantity of motion. What results here is a clearer picture of how cause and effect are related in the early part of the dynamics project.

The use of the notion of "equipollence" in the principle of equipollence relied on the development of a methodology by Leibniz and his forebears. This is the methodology of statics and the center of gravity. We have already discussed the central aspects of Leibniz's use of center of gravity and static methodologies in the previous chapters. To put a finer point on the issue, we can begin by pointing to what Leibniz called "an abuse" of statics (traditional mechanics) at least by 1685 in *Brevis demonstratio erroris memorabilis Cartesii*.[1] Now, this terminology of "abuse" arises in some later works by Leibniz but the point is illuminating for the origin of the dynamics project itself. What constitutes "abuse" is the treatment of the downward force responsible for equilibrium on a lever in terms of the conservation principle of motion. For example, in the work of near contemporaries like Wallis, the statical concept of weight and distance to fulcrum was held as equivalent to weight and speed in calculations concerning the center of gravity.[2] If we start by taking the explicit reason that Leibniz gives in his late 1690s supplement to the *Brevis demonstratio* for why the argument of *vis* from the law of the lever counts as abuse, we see that the key point here is that downward force only translates to Leibnizian *vis* in the entire range of motion that it engenders rather than merely the static equilibrium between moment downward forces. Why does the duration of motion matter for the laws of motion? We highlight this question in order to return to it later.

Another reason Leibniz gives for criticizing the abuse of statics in its extension into the theory of motion is that it is only a coincidence that the two cases, of the conservation of motion and the lever in equilibrium, coincide. It is a coincidence because there is no necessary causal relationship between static downward force and the quantity conserved in motion. Mere weight does not produce the effects of speed in freefall. Since the law of the lever expresses the symmetry between mass and distance to the fulcrum, it is merely the symmetry common to statics and the rules governing collision that such an equivocation was formulated. By reasoning from this intuition based on symmetry, the mechanists of this period sought to identify the symmetry produced by collision with the law of the lever. Faced with the same problem, Leibniz searched instead for a causal connection between center of gravity principles and the principle of motion.

Let us explain this difference with the use of Leibniz's canonical example.

---

[1] For a definitive comment on the abuse of *statics*, see *Essay de dynamique* (GM VI 218).

[2] See Wallis 1695, 682, 1670, 168.

The causal relationship between center of gravity and speed was developed in *De arcanis motus* but solidified in the later *De corporum concursu*. Now, in this latter text of 1678, Leibniz began his writing with the conception that the equipollence of cause and effect corresponded to the rectified Cartesian conservation of the quantity of motion (with direction). Hence, it should not be ignored that the principle of equipollence is compatible with the rectified Cartesian conservation principle. As Leibniz states in the *Brevis demonstratio* years later,

> Seeing that velocity and mass compensate for each other in the five common machines, a number of mathematicians have estimated the force of motion by the quantity of motion or the product of the body and its velocity. [...] This led Descartes, who held motive force and quantity of motion to be equivalent, to assert that God conserves the same quantity of motion in the world. (GM VI 117; L 296)

Indeed, Leibniz was himself susceptible of this same "abuse" of statics. The direct correspondence of the law of statics to the conservation of "force" was seen by Leibniz as a legitimate starting place. The only problem was that the Cartesians (and others) drew the wrong conclusion from statics, while Leibniz found a better path. Through the composition of *De corporum concursu*, Leibniz came to recognize and note the error that had guided his work up to that point.

> I now see where the error is to be found. The force in bodies should not be measured [*aestimanda est*] from speed and the size of bodies but from the height from which it falls. Hence the heights from which bodies fall are as [a proportion of] the square roots of the speeds in question. [...] Thus generally, the *vires* are in a ratio composed of the simple product of the bodies and the square of the speeds. (Leibniz 1994, 134)

From this, the equipollence between maximum speed and maximum heights is modeled on the center of gravity. It indeed follows from the symmetry of the law of the lever that the quantity of work is conserved in the raising of the bodies of different masses. We know retrospectively from the energy-work theorem that the law of the lever can be derived from this. However, this result cannot directly be translatable to the conservation of momentum in (elastic) collisions. When Leibniz construes his opponents (various Cartesians, Wallis, Wren, and others) as abusing statics, he thinks of them as identifying the speed of the falling bodies with their respective heights at equilibrium. If their speeds are not linearly proportional to their masses and heights, this results in an inequality which invalidates the universality of the law of statics. In this sense, the conservation of mv^2, based on the Galilean law of falling bodies, corresponds better to the law of statics than does the conservation of the quantity of motion.

The conservation of the quantity of motion is thus, in Leibniz's view, a result derived from a coincidence due to the correlated but non-causal relation between the symmetry operative in the law of the lever and the conservation of momentum in collision.

Here, the question bracketed earlier about the difference between a moment in the mechanical system and the extended duration of motion of that system becomes relevant. What is accidental in the "Cartesian" theory is the symmetry between bodies in equilibrium, in stasis, and the collision between two bodies. Now, this accidental correlation can be disturbed if we consider that the rate of falling bodies is a contingent feature of the universe. This acceleration is only, by definition, expressed in duration. In fact, Leibniz's argument in *De corporum concursu* took the contingency of the rate of falling bodies as a key part of his argument for the rejection of the Cartesian quantity of conservation. He notes, "Perhaps in another system of the world where speeds would have another relation to their heights, we would then require another measure of forces […] the laws of motion are nothing but the reason of divine will which assimilates effects to their causes to which the measure of things are subject" (Leibniz 1994, 134).

To distinguish the law of statics from the conservation of *vis*, we can imagine a world where the law of freefall were different, as Leibniz considered in *De corporum concursu*. However, we could also consider the case of a lever (where two bodies are balanced) raised to different heights without changing the law of falling bodies. The property of equilibrium on the lever, with bodies on different sides of the fulcrum, remains unchanged. From a Leibnizian perspective, what is changed in this raising is the absolute increase in *potentia*. It is this concept of *vis* qua *potentia* that was developed in this early period of the dynamics project. Now, according to the theory of the center of gravity, we consider the motion produced by a closed physical system (using modern terms) admitting of no external forces. With Torricelli's principle, a system cannot raise a body higher than its starting center of gravity. Using Leibniz's standard argument from this period, we understand that a pendulum with two bobs raised to a certain height cannot raise the center of gravity between the two bobs higher than its starting position. Of course, a pendulum can be examined at any point of its motion. What is required here is that the maximum height reached by the two pendulum bobs establishes a center of gravity that cannot be exceeded by any further motion. Because this center of gravity cannot be exceeded in the closed mechanical system, the two bobs in the pendulum constantly maintain the same amount of *potentia* distributed between the two masses translating between speed and a certain amount of achieved height. It obviously follows that maximal speed and maximal height in the system are proportional with respect to the *potentia* of the system, according to the equipollence of cause and effect. It also follows, from Torricelli's principle, that there can be no perpetual motion machine. Since some *potentia* is lost to friction in actual physical systems, the exchange between maximum height and maximum speed is always imperfect.

In changing the law of falling bodies or the arbitrary raising of the height of the lever, the law of equilibrium remains untouched just as any collision would be symmetrical given a closed physical system. The only difference here is between the speeds of the falling bodies. In this case of the arbitrary raising of the center of gravity the proportion of speed and distance to the fulcrum would differ, but since gravitation is constant, the relative center of gravity between the bodies would remain the same. The difference in this case is in the resulting speed from bodies raised to different heights. Hence, the "effect" of *vis* would be contingent on the role played by the law of falling bodies (and its modification) and the reciprocal relation between work and the terminal speed of a falling body. The result is that there is a causal relationship between the center of gravity and motion. However, it is not merely geometrical symmetry that grants this relation. If we arbitrarily raise the center of gravity, or if gravitation was systematically different, the symmetry of momentum could not reflect this difference. There is a causal relationship between height and motion in a way that is not accidental but rather a consequent of the law of falling bodies. In other words, the contingency of the law of falling bodies establishes a causal relation, and thus the actual nature of the law of falling bodies in this world provides the causal link between the conservation of *vis* and the capacity for motion (qua future effect).

What this analysis underlines is Leibniz's establishment of the crucial relationship in the early dynamics project between the contingency of the law of falling bodies and the equipollence of cause and effect. If cause is the capacity to produce future effect, then the problem raised in this episode at the beginning of the dynamics project is the structural translation of this relationship. That is, given the same theory of equipollence, different principles can hold. The measure of mv^2, derived from the law of falling bodies, grants a concrete interpretation of the theory of equipollence in this world.

5.3   Intensity and Extension in the Statical Methodology of the Dynamics

In the examination above, we have seen how Leibniz attempted to move beyond statics by introducing an absolute measure of *vis* that could capture a relationship between motion and work that eluded the standard statical methodology of the time. However, the theory of *vis* remains tied to statics. Even with a nascent theory of causation in germination, the theory of *vis* was a theory of *potentia*, an intensity to be exhausted or expended through motion. We shall explain this in terms of the causal relationship between conservation and motion.

As we have seen, the concept of *potentia* was developed by Leibniz through measuring maximal values of velocity and height in the case of the pendulum. In this case, the constraint of the center of gravity provides an invariant that allows a proportion for these two maximal values as well as a notion for absolute values of *vis* distinct from mere relative speeds. After the *De arcanis motus* (1676), this remains a cornerstone and constant feature of Leibniz's account. However, two different observations are crucial for understanding the limits of the concept of *potentia*. The first is that these mechanistic arguments are incapable of accounting for the distinction between inertial and non-inertial (accelerated) motion from the perspective of *vis*.[3] The second is that the proportion between the maximal values of height and speed are contingent on the actual acceleration of gravity. Let us treat these two different observations in turn.

---

[3] We note here, as in the second chapter, that inertial and non-inertial motion should not be confounded with inertial and non-inertial reference frames.

First, we should consider why the account of *potentia* from strictly mechanical principles is incapable of distinguishing between inertial and non-inertial (accelerated) motion. The simple answer to this is that *potentia* is a structural property derived from maximum values. With the notion of "future effect," the maximum work of the system is put into proportion with the maximum velocity of the system, placing Leibniz among the many seventeenth century forebears of the virtual work principle and the calculus of variations. However, for Leibniz, the measure of *vis* reveals only absolute differences, for example pendulum swings from different heights or an arbitrarily raised lever, can only be distinguished for these maximum values. What escapes this method is *how* those values are produced. Of course, Leibniz was deeply aware of the fact that acceleration was involved and even provided an account of this acceleration in the form of the speed of pendulum motion and the uniform acceleration of falling bodies. However, this account was not part of the initial presentations of the concept of *vis*.

Indeed, in the initial public presentation of the initial findings of the dynamics project, organized as a critique of Cartesianism, the *Brevis demonstratio*, the response from the Cartesian camp was very clearly one of taking Leibniz to task for not considering how the maximum values that established the measure of *vis* were accomplished. That is, the temporal account of the transformation that brings a physical system from a state of maximum speed and to that of maximum height. The Cartesian Abbé de Catelan's criticism of Leibniz was that, within the account of the time of that transformation, there simply could not be a proper measure of the quantity of motion for that system. As will certainly recur in Leibniz's many varied correspondences, Leibniz and Catelan misunderstand each other fundamentally (GP III 40–42; GP III 51–55).[4] Catelan points out that to properly refute the conservation of the quantity of motion, it is not only the downward fall but also the upward rise that needs to be taken into account. This Leibniz had not done. Leibniz, on the other hand, saw that it was enough to argue that the quantity of motion does not describe how a physical system produces work. In his response to Catelan, Leibniz promptly sidelines the problem of time and moves to treat the tautochronicity of curves (geometrical acceleration) resulting from gravitational force.

From this exchange concerning the *Brevis demonstratio*, it is sufficient to grasp Leibniz's disregard for the factor of time in his concept of *vis* qua *potentia*. Indeed, even in his late 1690s supplement to the *Brevis demonstratio*, he maintained that, with respect to the problem of work, time is of no consequence to the demonstration itself. "Universally, power is measured by effect and not in time; time can be varied by external circumstances" (GM VI 122). With time rendered static through "future effect," *potentia* is derived for a proportion that is indifferent to the accelerations needed to generate the transformation between height and speed. This amounts to a limited methodology for the treatment of the causes of motion, one that is based on the proportional ordering of *potentia* and its effects compared across different systems. Insofar as the dimension of time is reduced to "future effect," the proportion between *vis* and motion can only be grasped when there is a comparison between

---

[4] See also Denis Papin 1689, 183–189.

absolute magnitudes of *potentia* in a system. Hence, the causality of *potentia* can only be understood externally as the achievement of a final cause, the conservation of a magnitude responsible for harmonizing different systems of *potentia* and effects.

Hence understanding *potentia* as a systematic property is limited in its capacity to produce a theory of the cause of motion. *Vis* qua *potentia* is an extension of statics but one that still remains too tied to it. Its capacity to reveal causation is hence limited to conservation of the proportion h$\propto$v^2 between physical systems of different absolute magnitudes of *potentia*.

Second, the maximal values of height and speed are contingent on the actual acceleration of gravity. In the immediately preceding discussion we indicated the limitations of the concept of *potentia*, here we are dealing with the modality of measuring *vis* qua *potentia* (Leibniz 1994, 134). The proportion h$\propto$v^2 is a result of the fact that gravity accelerates bodies quadratically. Hence, the equipollence of cause and effect provides the formal placeholders that are filled by the actual values contingently determined by the acceleration of gravity on Earth. Leibniz explicitly provides for the possibility of a different rate of acceleration. Although Leibniz does not elaborate here, it is abstractly possible that gravity is non-accelerative and that in another world a body falls at constant velocity. As such, the difference between the measure of *potentia* qua *vis* through mv^2 is only a matter of the teleological harmony at work in the actual world. Leibniz will clearly state this in a letter of 22 November/2 December 1679 to Christian Philipp that the theory of force "depends" on final causation (A II 1, 767). In other words, the problem is again that the nature of *vis*, qua final cause, would result in an entirely external or extrinsic determination of motion. This renders the quarrel between Leibniz and the Cartesians into a dispute merely over the measure corresponding to the theory of conservation.

Of course, Leibniz saw that the quarrel over the measure of conservation is itself indicative of a deeper metaphysical disagreement. Through this quarrel, Leibniz was able to qualify the laws of motion as contingent on the creation of the actual world. In this sense, even if Leibniz somehow gave up on the attempt to establish his alternative to the Cartesian laws of motion, he would still be able to argue for a distinction between mere geometrical relations and the "dynamical" relations involved in physical reality. In this sense, the importance of the alternative measure of *vis* qua mv^2 remains that of final causation. That is, the actual laws of motion reflect the wisdom of the creator-architect who chose between different geometrical relations for the relations that actually obtain in physical reality.

From the two points analyzed above, we see how the initial conception of *vis* qua *potentia*, through their explanatory limitations, constituted cause as only final cause. The problem, as I have argued, is that the causal nature of *vis* qua *potentia* remains an external principle of motion governing the teleology or teleonomy of motion rather than an inherent or intrinsic principle of motion. Hence, the theory of *potentia* is limited precisely because it cannot stand as an immanent cause of motion.

One way to understand this limitation of the dynamics project at this point in its development is to understand the project as a methodology for comparing causes and effect along the lines of the equipollence of cause and effect. What Leibniz attempted to provide at this point was not a theory that would give an account of the causes themselves but only an external quantitative means to evaluate *potentia* across systems. Hence what we have established in this section is that the limitation of the concept's development during this period pertains to the analysis of causation within the early part of the dynamics. Indeed, the adoption of the equipollence of cause and effect as a architectonic principle was not enough to arrive at the conservation of mv^2. It was this methodology however that led Leibniz to privilege the quantity mv^2 in his "reform" of mechanics in the 1678 writings. As such, what we shall see is that there was a development of the concept of "cause" that attempted to move beyond the inherited methodology as he entered more deeply into the dynamics project.

5.4 *Potentia* as an "Incorporeal" Principle

As we have argued in the precious section, the early period of the dynamics saw the development of a methodology that remained at the heart of the development of the project until the end. The "axiomatic" equipollence of cause and effect remained an explicit justification of the dynamics even in the latest manuscripts. Yet this methodology could not satisfy a central desideratum of the dynamics project, a problem that had accompanied Leibniz across a long period in his philosophical development. This was the search for an immanent principle of motion within physical reality.

Let me set the stage briefly with a quick summary. Leibniz's earliest publications were the twin texts *Theoria motus abstracti* and *Theoria motus concreti* (*Hypothesis physica nova*), two short treatises dedicated to elaborating a physical theory capable of synthesizing the mechanistic methodologies of the moderns (in terms of physical explanation) while leaving enough room for the traditional metaphysics of Aristotelian causes. Without being either dogmatically mechanistic or faithfully Scholastic, the ecumenically disposed Leibniz turned the domain of physical theory into an arena to address his much deeper concerns about metaphysical and theological doctrines. What accompanied Leibniz from the writings on physical theory of the late 1660s through to the origins of the dynamics was the attempt to discover an immanent principle of action within physical reality. This, of course, took several forms.

Now, without discussing the many different forms that this immanent principle of action took within the evolution of Leibniz's physical theory across the years, what is crucial for us is to see the role that this metaphysical intention played within the formation of the dynamics project. To grasp this, we should see that the general role of explanation played by mechanical processes remains largely unchanged between Leibniz's youth and maturation. In his early letters of the late 1660s to his mentor Thomasius, Leibniz argued for the compatibility of Aristotelian causes and the new mechanistic philosophy by granting the role of explanation in terms of size, shape, and motion (A II 1, 25; L 94). Now, we know that one of the key factors of the early development of the dynamics project surrounding the *Brevis demonstratio* was the rejection of size, shape, and motion as sufficient for a complete physical theory. Nonetheless, in mature texts like *Specimen dynamicum*, Leibniz was able to affirm that, although the causes of motion were only accessible through "metaphysical reflection," everything was still explainable based on geometrical relations established by size, shape, and motion. Hence, what is "insufficient" about mere mechanistic explanation was, as we have argued, the choice of the mechanical principles among a range of possible natural laws. This, we have argued, safeguards the interpretation of the dynamics as balancing the cause of motion between efficient mechanical causation and final causation based on conservation. There is, however, an important missing piece to this theory of causality.

In the period of Leibniz's youth that we have now been considering, the ecumenicalism of the early physical theory was based on a distinction between efficient and formal causation rather than a distinction between efficient and final causation. As he argues in his 1668–1669 *Confessio naturae contra atheistas*,

> At the beginning I readily admitted that we must agree with those contemporary philosophers who have revived Democritus and Epicurus [...] that so far as can be done everything should be derived from the nature of body and its primary qualities –magnitude, figure, and motion [...] But what if I should demonstrate that the origin of these very primary qualities themselves cannot be found in the essence of body? Then indeed, I hope, those naturalists will admit that body is not self-sufficient and cannot subsist without an incorporeal principle. (A VI 1, 489–490; L 110)

Hence, the division that allowed Leibniz to balance the moderns against the Scholastics was a division between the corporeal or phenomenal and the incorporeal. It is this "incorporeal" principle within the corporeal that remained a constant unfolding metaphysical theme throughout the Paris period. The establishment of the "rehabilitation" of substantial forms, the key metaphysical program of Leibniz's so-called middle period, would also be based on how only a theory of form could individuate bodies and motion.

The convergence of Leibniz's mechanical studies and metaphysics during the middle period has been the topic of many scholarly writings in recent decades, so I shall leave the broader themes and controversies aside. What concerns us here is only to clarify the sense in which substantial form is important to the constitution of the dynamics project. The point in question, a theme that connects Leibniz's earliest writings on physical theory and the last writings on the dynamics, is Leibniz's search for an inherent principle of action in bodies. Thus the progressive crafting of a physical theory developed by Leibniz even before his departure to Paris was guided by the metaphysical intention of establishing a demonstration that, "the substance or nature of body, also in accord with the Aristotelian definition, is thus the principle of motion (for there is no absolute rest in bodies); where the principle of movement or substance of body is [taken] apart from extension..." (A II 1, 281). By 1679, this program found expression within the newly formed dynamics project. In the *Conspectus libelli elementorum physicae*, written soon after the *De corporum concursu*, Leibniz argues that it is the equivalence of cause and effect that allows for the demonstration of this "incorporeal principle":

> There follows now a discussion of incorporeal things. Certain things take place in body which cannot be explained from the necessity of matter alone. Such are the laws of motion, which depend upon the metaphysical principle of the equality of cause and effect. Therefore, we must deal here with the soul and show that all things are animated. Without soul or form of some kind, a body would have no being, because no part of it can be designated which does not in turn consist of more parts. Thus nothing could be designated in a body which could be called "this thing," or a unity. (A VI 4, 1988; Garber 2009, 51)

We should notice how Leibniz argues here. He moves from the irreducibility of the laws of motion to geometrical necessity, an appeal to the need for final causes, to an argument for the incorporeal principle as a "soul" animating the behavior of bodies and granting its "being." From what we have argued in the previous section, this is a significant but unwarranted leap. Now, we can also see why Leibniz argues in this way. The problem of soul or the "incorporeal principle" is tied in this passage to the individuation of bodies from material parts. If we take the constitution of bodies from material parts to be granted its being (its unity) through a substantial form of corporeal substances, then it appears that the constitution of motion from its continuous parts can be granted its being (its unity) through a substantial form of motion. This analogy certainly was the intention behind Leibniz's claim and similar ones that we can find throughout the development of the dynamics project. The only problem here is that the dynamics project remained unable to supply just such an account for the formal cause of motion. That is, the initial results of the dynamics were suitable for claiming an "incorporeal principle" through the methodology of the equipollence of cause and effect but remained at a distance from providing an inherent account of the action of bodies that produces the magnitudes expressive of such a final cause.

What we have tried to show here is the difficult tension in the dynamics project between the metaphysical intention to treat *vis* as an incorporeal principle and the foundational methodology of the dynamics based on a concept of *vis* as a conservation quantity. Of course, for the metaphysical implication of the early dynamics project, the contingency of the rate of falling bodies is crucial to the argument about the difference between mere corporeal motion and its underlying "incorporeal principle." Yet, as we have argued, since the cause of motion is understood through *potentia*, this serves only as a placeholder for a certain magnitude of the capacity for motion. Hence, a difference in gravitational acceleration would mean a different result for the proportions of work and velocity, a different proportion between absolute measures of *potentia*, and ultimately a different relation between cause and effect. It is precisely this difference that makes the "incorporeal" principle incorporeal and essential to the metaphysical implications of the dynamics project at this point.

Even if the concept of *potentia* represents the capacity to accomplish work, it says nothing about the time involved in the accomplishment of this work. As such the effect of *vis*, the effect of a cause, is contingent upon a key proportion that is not itself a mechanical relation. The contingency of Galileo's law of falling bodies is then a conspicuous exception to the mechanical relations between bodies. Despite the usefulness of this principle of the equipollence of cause and effect in concretizing his rehabilitation of substantial forms, it remained insufficient for a theory of formal causation that would allow Leibniz to successfully identity an "incorporeal principle" of corporeal motion. This required the dynamics to move beyond the concept of *potentia*.

5.5   *Actio* and the Immanence of *Vis*

In the previous sections, we have argued for the importance of the Galilean law of falling bodies for the development of the dynamics. Not only did this serve to provide Leibniz with a conservation quantity which allowed him to engage in the exploration of a mechanical theory opposed to the Cartesians but it also allowed him to draw out some of the metaphysical implications concerning the contingency of the laws of motion from the emerging dynamics project.

As we have already mentioned several times, the dynamics was only named in 1689, and it was in this period that Leibniz composed two long texts aimed at providing the substantial treatment of physical theory that he had been putting off for quite some time. It was also in the context of these writings that Leibniz would replace the identity between *potentia* and *vis* with the term *actio*. *Actio* finally becomes the formal cause that we have been searching for in Leibniz's dynamics. What is interesting here is that, in doing so, what shifts in Leibniz's dynamics project is the move away from the reliance on Galileo's law of falling bodies taken as a given. We shall examine how this turn to a principle of formal causation transforms the understanding of the principle of the equipollence of cause and effect as an "axiom."

In the preface to the *Dynamica*, Leibniz provides an argument that explicitly turns the concept of *vis* away from the assumption of Galileo's law of falling bodies. The argument involves a theory of formal effect that incorporates the concept of work within a single rectilinear motion. The argument signals a shift in Leibniz's conception of *vis* because it does not involve the usual and perhaps more obvious accounts of *potentia* involving the work of raising a body to a certain height. Part of the question to keep in mind here is, as we have touched on in previous chapters, whether the measure of *vis* follows from the law of falling bodies (taken as a given assumption) or whether *vis*, as an immanent principle of corporeal motion, is in fact responsible for the measure of gravitational acceleration. With the aim of evaluating *vis* from the perspective of the immanent activity of bodies in mind, we recall the argument for the determination of *actio*, discussed in earlier chapters. Leibniz argues in the *Dynamica* that,

> The action bringing about the double [effect] in a single [unit of] time is twice the action of bringing about the double [effect] in double the time.
>
> The action bringing about the double [effect] in double the time is twice the action bringing about the single [effect] in a single [unit of] time.

Therefore, the action bringing about the double [effect] in a single [unit of] time is four times the action bringing about the single [effect] in a single [unit of] time. (GM VI 291–292)

In a slightly different exposition of this argument in the *Essay de dynamique* a decade later, Leibniz notes that,

To the end that motive action can be measured, we first need to measure the formal effect of motion. This formal or essential effect of motion consists in what is changed by motion, that is to say the quantity of mass that is transferred and the space or length through which this mass is transferred. [...] Now it will be easier to understand what is motive action: we need not only to measure the formal effect that it produces but also the vigor or velocity through which it produces the effect. (GM VI 220–221)

What is the theory of effect ushered in by the system of action? From the argument here, it is clear that the concept of action does not require the theory of future effect even if such a manner of speaking remains appropriate to the dynamics. The center of the theory shifts from a theory of intensity and its expenditure to the production of phenomena. Recall that action is understood as the product of formal effect and velocity. As formal effect increases, velocity diminishes with respect to this increase.

$$a = ms \cdot v$$

$$a / \Delta v \propto \Delta ms$$

The conservation of *vis* and the invariance of action thus amount to the same quantitative magnitude but differ in theoretical content. The invariance of action is based on the temporal evolution between effect in time and state of motion. The issue here is not simply that time is included in the account of action but rather that time plays a completely different role. The time of *potentia*, understood as the time of future effect, is the time for the accomplishment of some work. The time of action is an independent time, an independent continuous framework in which *vis* acts. The conservation quantity mv^2 regulates the relation between on-going formal effect and the evolving state of velocity. With time taken into account, action governs the relation between formal effect and the state of velocity through the constant invariant mv^2 at every point in this evolution. This more analytic account of the relation between *vis* and motion was previously unavailable. And indeed this was the limitation picked up by Leibniz's Cartesian critics like Catelan, Malebranche, and Papin.

With the temporal dimension included in Leibniz's thinking, the dynamics moves toward a new conception of cause and effect. In the earlier conception, the time of motion translates an intensity (*potentia*) into extended motion. When this intensity is expended, cause coincides with effect. In the later conception, the notion of *potentia* as an intensity that is exhausted in time is replaced with an action of *vis* in time that constantly and immediately coincides cause and effect. This is because time is no longer the time for the completion of a motion within an equilibrium but rather the immediate expression of an invariance captured by a proportion between extension and intensity. *Vis* constantly produces motion, but it produces motion according to its state of velocity. This state of velocity is dependent on its relation with its quantity of formal effect.

The coinciding of cause and effect in the concept of action is due to a key conceptual shift. Moving the static relationship between maximal quantities between motion and work, what is at stake in the concept of *actio* concerns the translation between these two quantities. Indeed, Leibniz turns to focus on the causal production of physical reality. It is this relationship that is formalized in the concept of action. Action captures the temporal evolution of this relationship that goes beyond the relation established by equipollence. With this, the equipollence of cause and effect is almost entirely drawn away from its initial context. To replace this, the principle of equipollence becomes the constant identity (in time) between the cause of motion and the proportion between the two magnitudes of formal effect and velocity.

The claim of this chapter and also the central claim of this book as a whole is thus dependent on a basic shift in Leibniz's theory. Dynamical causality becomes structural causality. The key transformation here is that, while *vis* could still be understood as an entirely mechanical relation (governed by final cause), action revises these relations. Hence, while mv^2, the proportion realized in mechanical transformation, could still be understood as the actualization of a magnitude through motion (a final cause), the mv^2 of action is realized in the immediate cause of motion. The effect produced by the cause (*vis*) is not only the state of motion but also formal effect. It is this immediacy between cause and effect that allows us to deem this a structural cause rather than either an operative cause (Newtonian force) or a mediated final cause, the intensive power to produce future effects.

Hence a conservation quantity, such as mv^2, may indicate an incorporeal principle or even a final cause in the causation of physical motion, but it is not an inherent property of body except in the sense that final cause indicates a non-mechanical determination of the laws of motion. Here the concept of *actio* allows Leibniz to transform a final cause into a formal cause within corporeal motion without disrupting the theoretical edifice that he had built up through the dynamics project. What, however, allows Leibniz to make such a theoretical step? Let us return to our earlier discussion of the equipollence of cause and effect and the question of the "abuse" of statics.

5.6   *Vis* qua *Actio*

Understanding the evolution of Leibniz's method for the analysis of motion is important for understanding the concept of *actio*. Just as the concept of *actio* was an attempt to stretch the limits of the dynamics beyond its statical methodology, so was the development of the distinction between *vis* expressed as dead and living *vires*. Note here that the same conservation quantity is at work in these two different expressions. With this, the reliance of mv^2 on gravitational acceleration is untethered, and *vis* can find a concrete and independent status as the cause of motion through *actio*. How does this occur?

Dead and living *vires* are distinguished through time. Dead force is momentaneous motion, and living *vis* is the *vis* of a motion that has been extended. Now, then, the difference here, according to the concept of action, would be the difference between the action of a body considered according to an extended motion and the action of a body momentaneously. Since dead *vis* is measured according to the impulse to move, it represents the state of motion in a body. This is one of the two factors within the account of *actio*. Recall that the measure of *actio* is the product of the formal effect and the tendency to move. Dead *vis* is only the momentaneous tendency to move understood as the effect of *vis* within the evolution of motion. In the supplement (1697) to the *Brevis demonstratio*, Leibniz illustrates this difference of a factor in a geometrical way:

> For living power is to dead power, or impetus (actual velocity) is to conatus, as a line is to a point or a plane is to a line. Just as two circles are not proportional to their diameters, so the living forces of equal bodies are not proportional to their velocities but the square of their velocities. (GM VI 121, L 299)

Dead and living *potentia* (or *vis*) is geometrically distinguished through dimension. However, as it is clear from this geometrical analogy between diameter to area of circles, the one is not the cause of the other. As we have argued in the previous chapters, this analytical tool, moving Leibniz's method beyond statics, does not stand in for an account of causality. The question here is how this dimensional difference is bridged between moments of motion and the extended motion caused by *vis* (*viva*).

Through the adoption of the concept of *actio*, Leibniz also distances himself from the earlier methodological need for the assumption of the Galilean law of falling bodies. In the scholium that follows the passage from the "Preliminary specimen" of the *Dynamica* examined previously, Leibniz notes that,

> [W]e already not only have a remarkable agreement among truths, but also in a new way is opened for demonstrating Galileo's propositions about the motion of heavy bodies without the hypothesis he had to use, namely, that in their uniform accelerated motion, heavy bodies acquire equal increments of velocities in equal times. For this very fact, as well as the lemma assumed above, can be derived from our fourth demonstration, which does not depend on them as assumptions. (GM VI 292; AG 111)

Now, what the reliance on the law of falling bodies guaranteed is a quadratic proportion ($v^2 \propto h$) between maximum height and maximum speed (within the operation of a mechanical system). As we saw in the last chapter, this kind of methodology remains tethered to the statical methodology inherited by Leibniz. However, Leibniz also attempted to move beyond the statical methodology. The movement away from a methodology tethered to statics certainly has little in common with those he accused of an "abuse" of statics. It is aimed at developing a theory of the translation of *vis* to motion that could account for grounding the needed Galilean law of falling bodies and hence for qualifying the contingency of the theory of *vis*. What this implies is the priority of the theory of *actio* and its concretization of the measure of *vis* over the law of falling bodies which Leibniz had frequently taken as given. With this new foundational account, we can maintain that, in *a posteriori* experiments, the law of falling bodies, is assumed to provide a measure of the exchange between the *potentia* of the maximum raised heights and the maximum speeds of that system. However, the theory of *actio* presents the *a priori* causal ground for the quadratic ratio ($v^2 \propto h$) responsible for generating the magnitudes involved in gravitational acceleration. Hence regardless of the mechanical vortex theory of how gravity behaves on this planet or another, what underlies it is the fundamental and universal law of action. In this, the problem of how *vis* translates into motion becomes a key issue in the *Dynamica*. It is by questioning the direct relationship between *vis* and motion in time that a new theory of cause and effect arises. In contrast, the theory of *vis* understood through the conservation of *potentia* is indirect. The earlier theory of *potentia* is essentially indifferent to the *means* by which maximal heights and velocities are achieved. Since the quantity conserved in motion is mv^2, the higher dimensionality seems to imply that some degree of acceleration must be required to understand how mv^2 translates into motion.

This is the impression we have from examining Leibniz's arguments that aimed to show the difference between *vis* viva and *vis* mortua, for example, by pointing to the proportion between diameter and area of a circle. It is only when we consider how *vis* is also acting in non-accelerative motions that we understand the universality of this concept within Leibniz's dynamics.

For non-accelerated motion, a body moves at a constant rate (with respect to the inertial frame). If $s/t = v$ is speed, then after moving for a certain time (t), the body is displaced by a magnitude of s. Hence the effect of inertial motion, translated in time, is the distance traversed by the motion in a certain amount of time. Yet since the motion is inertial, the state of the body is still moving at speed v. The action of the motion is thus the product of the effect and the state of motion of the body $a = m \cdot s \cdot (s/t)$ (GM IV 388–389). This is how Leibniz reasons that two factors of speed must be considered in any motion. The actual account of action also includes mass hence $a/t = m(s/t) \cdot v$, or $a/t = m(s/t) \cdot (s/t)$.

From a contemporary standpoint, this is rather trivial because the conservation of energy-work is conservation for systems involving both inertial and non-inertial motion. But this signals a conceptual innovation for Leibniz in the analysis of the translation of *vis* to motion. *Vis* is no longer merely concerned with a general conservation property of motion. Rather, Leibniz begins to conceive of *vis* as the action of a physical system in time. This action of course is given a general rule in the theory of *actio*. If this holds for inertial systems, it certainly holds for the more obvious case of accelerated motion.

For an accelerated body moving at v at time t, the change of time will change the magnitude of v quadratically. Hence for $a/t = m(s/t) \cdot (s/t)$, or $a/t = mv^2$ as time increases, v changes quadratically. This reflects the accomplishment of work understood on par with the constant of action over time.

As we have examined above, the relation between "future work" is replaced with the immediacy of *actio* in the determination of the causality of motion split between two factors, formal cause over time and the momentaneous speed of a body at that time. We can draw two clear implications here.

First, the concept of action is a universal determination of motion. It not only applies across inertial and non-inertial motion but also provides the means of distinguishing between the two. This means that, regardless of the acceleration of gravity, the action of *vis* correctly determines the relation between the magnitude of motion achieved by the motion in time and momentaneous velocity. Applying across inertial and accelerated motion, the universality of *vis* qua action, over and above *vis* qua *potentia*, grounds the claim in the dynamics as a theory that provides a universal account of motion.

Second, the concept of action presents a theory of the translation of *vis* into motion. Here, *vis* is immediately the cause of motion: *vis* understood as *actio* in the distribution of *potentia* in time. If we take this as a magnitude, the quantity of *actio* is expressed as a distribution of the effect of speed over time. This is the most immediate interpretation of *actio*. That is, if *potentia* is mv^2, we need to understand how this quantity concretely unfolds in the time of a body in motion. With the theory of *actio*, Leibniz provides us with a concept of the translation between cause and effect. It is in this that we see how the dynamics was actually brought to its full maturation.

5.7 Dynamical Causation

In the maturation of the dynamics, the concept of causality that Leibniz wished to capture with the concept of *vis* should be understood as a structural causality. This shift from *vis* as *potentia* to *vis* as action can be seen mostly from the transformation of the concept of effect. As we examined above, the relationship between cause and effect shifts from the model of the expenditure of a certain mechanical power to the action of a body to bring about a proportion between the formal effect of the physical system and its state of motion. As interpreters like Duchesneau point out, the dynamics shifts in 1689 from the treatment of the relationship between two terms *potentia* and speed to a relationship between three terms, *actio*, speed, and formal effect (Duchesneau 1994, 147–262, 1998, 77–109). This implies that the quantity of conservation is not simply a distant final cause that is eventually achieved by corporeal motion as a kind of "future effect" but rather an immediate aspect of any corporeal motion at every moment in time. The *actio* of a moving body thus constantly determines the distribution of *vis* in space and time as a motion evolves. This development of two factors allows us not only to distinguish between inertial and accelerated systems but also holds for both these cases, allowing Leibniz to demonstrate the universality of the conservation of *vis* rather than simply assert it on the basis of mere conservation of *potentia*.

Leibniz's mature concept of dynamical causation is thus a relation between *vis* and phenomena. What is caused is not a certain magnitude that will be realized as "future effect" but rather the momentaneous properties of a body or physical systems that are actual at every moment of time. The distribution of *vis* in space and time or the translation of *vis* into motion renders cause and effect as immediately coinciding magnitudes. We can grasp this from two perspectives. The most obvious perspective is the intrinsic or immanent perspective. A moving body is imbued with an agency that was highly ambiguous in the earlier documents of the dynamics. The active capacity of *vis* is not only to achieve some future effect governed by an abstract and distant conservation principle qua final cause. Rather the *actio* of a moving body continuously produces a proportion between formal effect and state of motion. The body acts in a way that is not simply a spatial displacement or change of place. Rather the body acts by translating a conserved quantity into the phenomenon of formal effect and state of motion. In this way, what is immanently caused by the *vis* of a body or a physical system is the phenomena (in time) of that body or system. In other words, cause and effect are equipollent because of the immanent identity of cause qua action and effect qua phenomenon. *Actio* understood immanently is thus a transcendental principle for the possibility of motion. All motion is action of a body with respect to formal effect and state of motion.

What is less obvious from the concept of *actio* is the extrinsic perspective, since *actio* most immediately refers to the inherent action of a moving body in time. However, if we take a perspective that is not concerned with the intrinsic action of a moving body and only concerned with the conservation of the magnitude of *vis* (or *potentia*), we find that the concept of *actio* also significantly revises the concept of causation. We have argued that the equipollence of cause and effect was the methodology for comparing different absolute measures of *vires* in different systems. This is an extrinsic perspective because the comparison of effects made through the comparison of accomplished work was a means to understand *vis* externally through the comparison of different physical systems with different absolute magnitudes of *vires*. This was what made it originally possible for Leibniz to determine the quantity $h \propto v^2$. Systems with different maximum speeds indicate a magnitude of *vis* that differ quadratically. The methodology here is extrinsic because it did not account for how this difference of dimensionality between speed and *vis* was engendered. Rather, it relied on the assumption of the Galilean law of falling bodies. However, with the new method of *actio*, the extrinsic perspective, even without asking about how this proportion is engendered, determines absolute properties (magnitudes) of *vis* by comparing the product of the quantities of formal effect and speed in a syllogistic way. The need for relying on the quadratic relation between time and speed offered by the Galilean law was undone. Since the account of *actio* applies across inertial and non-inertial motion, Leibniz could show that his theory does not rely on the contingent nature of the rate of gravitational acceleration. Even if gravitation were not accelerative in a possible universe, though falling bodies would look very different, the theory of action would still hold. From an extrinsic perspective, the cause of the conservation of *vis* would be due to *actio* qua proportion between the change of formal effect and speed in time.

This perspective allows Leibniz to understand the analytical distinction between dead and living *vires* in a different light. Although we have already touched on this, we are now able to grasp the important role that this distinction plays in the dynamics. The difference between dead and living forces is extrinsic because it only considers the magnitude of motion involved in physical phenomenon. The difference in dimensionality, even if there is no precise way of accounting for this difference, means that the analysis of motion (inertial, rotational, or accelerative) into rectilinear proto-vectors is based on an overarching structure of the relation between states of motion (momentaneous speeds) and its place within a sequence of states brought about by the action of a body. Hence, from an external point of view, a body in motion is simply a sequence of positions and states of motion (speed). This is the concrete empirical effect of *actio*, the continuous production of physical phenomena.

From these two points, intrinsic and extrinsic, we can begin to appreciate how the dynamics brought about the achievement of Leibniz's agenda for the metaphysics of motion. This metaphysical agenda is two fold. It provides the phenomenon of motion with underlying unity and identity. The metaphysical problem of confronting the reality of motion, was, for Leibniz, that nothing in extension itself could guarantee its underlying reality. This means that nothing on the level of motion and extension could offer a principle of unity or identity for a moving body. In the development of the concept of action, such a principle of both unity and identity is finally achieved.

The first aspect of this agenda can be understood under the theme of unity. Retrospectively speaking, Leibniz's earlier theory of *vis* relied on an external view in order to grant reality to corporeal motion. This relied on the capacity of a body to accomplish a future effect in time. The sequence of phenomena that takes place is structurally united through a conservation principle. It is this unity through future effect that ties the sequence of locomotive phenomena into one coherent motion. The famous "maxim" that Leibniz formulates in his correspondence with Arnauld that "what is not truly *one* being is not truly a *being* either" is only applicable to an aggregation of discrete spatially distributed parts but not to dynamic evolution of temporal parts (A II 2, 186; GP II 97). With the conservation principle, *vis* is able to unite locomotive effects as expressions of a single entity. However, this unity remains ambiguous because the time needed to accomplish work could not be addressed within the framework of the conservation of *vis* as a structural property. With the development of the concept of *actio*, conservation does not require the notion of a "future effect". Rather what is conserved is present at every point in the temporal evolution of a motion. Through this maturation, the sequence of locomotive phenomena corresponds in total and in part to the unity that conservation grants motion. If the reality of motion is granted through the unity of motion, the move towards *actio* provides a concept of unity that is operative at every moment and across a temporal evolution.

We should add that Leibniz did not have a fully developed theory of physical phenomena before the theory of *actio*. From the external perspective of conservation, phenomenon qua effect is simply the accomplishment of a certain quantity of work. The temporality of physical phenomena was essentially unaccounted for except as a factor of velocity. It is only with the explicit treatment of time in the proportional relation between formal effect and speed that the continuous sequence of phenomena becomes the object of Leibniz's inquiry. From this, we can say that what is caused by *vis* is the unity of a sequence of locomotive phenomena. More precisely, the *unity* of a locomotive phenomenon has *vis* as cause.

The second aspect of the metaphysical agenda is the theme of identity. Now, identity and unity may indeed refer to the same thing but there is a subtle difference in *sense*.[5] A sequence of phenomena is united insofar as it is structurally one (an aggregate of parts), but its identity requires the immanence of cause responsible for this unity (an underlying *nature*). The concept of conservation is, by itself, unable to supply a concept of identity. As a principle of final cause, we can say that the divine always conserves the quantity of *vis* in the universe or a physical system. But this conservation can occur extrinsically or intrinsically. Since conservation is a final cause, this concept cannot, by itself, explain the means by which the conservation is realized except through purely extrinsic (phenomenal-mechanical) means. Hence, the concept of the intrinsic action of a body, *actio*, provides the dynamics with a theory of the identity of a physical system that determines the individual qualities of one physical system distinguished from another. *Actio*, for instance, can allow us to determine the difference between an inertial motion from an accelerative one. From the perspective of mere individuation through conservation this difference is reducible to the same thing. If *actio* determines the distribution of *vis* through a proportion between formal effect and speed, it places the realization of the final cause of conservation within the agency of a moving body or system. This grounds the reality of motion through an identity which is intrinsic to the action of a moving body.

With the realization of the twin metaphysical agenda in the theory of *actio*, we see how the grounding of the dynamics project shifts from mere conservation to a metaphysical doctrine of the inherent action of a body. The question here is then how to recognize a theory of structural causality once the structural property of conservation has faded into the background.

Let us consider this question from the perspective of the values of *vis* in different physical systems. Now, the conservation of *vis* was developed out of a proportion between work and speed. But of course this conservation will be identical between physical systems taken as unities. A partial physical system, taken as closed, expresses the same proportion between work and velocity as the larger physical system in which it is embedded. The magnitude of *potentia* is simply a constant that derives from the transformation of motion of all the bodies in the system into work

---

[5] This reflection on identity and unity strictly refers to the metaphysical consequences internal to the development of the dynamics. There is no intention to understand this as superseding the many other reflections on the metaphysics of unity, identity and, simplicity that characterizes Leibniz's profound and complex metaphysical itinerary. Of course, the role that the dynamics plays in this itinerary should not be underestimated.

in time. However, the absolute magnitude of *vis* is greater in the larger system than in the partial system.

It follows rather naturally that a partial system has a magnitude of *potentia* that is smaller than the magnitude of a larger system. The effect here will be a greater quantity of velocity and, proportionally, a greater quantity of work. Hence *potentia* is not scale-invariant, the measure of *potentia* in a smaller system is strictly lesser than that of a greater system. The quantity of *actio* insofar as it reproduces the quantity of *vis* qua *potentia* is also not scale-invariant. Indeed, the capacity for producing future effect is what allows us to distinguish different absolute values of cause and effect qua *potentia* and put them into proportions as Leibniz does syllogistically. The concept of *actio* however allows us to grasp these proportions synchronically at every state of the evolution of a physical system. Even if Leibniz will speak of different absolute measures of the capacity for actio in different systems of work, *actio*, operating as the proportion between accomplished work and the aggregate state of motion of a physical system provides a scale-invariant proportion between formal effect and state of velocity. The difference here concerns a different concept of causation. Since *potentia* is based on the model of exhaustion, the magnitude of effects measures the proportions between the *potentia* of different systems. If *actio* is based on the continuous production of phenomena, the magnitude of effects is relative to the immanent distribution of values of the system itself. The concept of *actio* still relies on absolute differences in the magnitude of *potentia* qua *vis* in different physical systems but provides a scale-invariant proportion between cause and motion since it governs the synchronic translation of causes into effects rather than the comparison between magnitudes of accomplished effects across greater and lesser scales.

In this sense, *actio* realizes an important feature of the dynamics project from a physical and metaphysical perspective. *Actio* provides an immanent structural principle for the cause of motion that satisfies the need for a methodology that distinguishes between different scales of *potentia* across different physical systems but also accounts for the "dynamical" causation of extended motion within a physical system. This significantly changes the meaning of the equipollence of cause and effect. Rather than an external comparison between magnitudes of *potentia* between physical systems, it shows how the immanent equipollence of cause can translate into extended effects that coincide structurally and immanently within every scale of the totality of the physical world.

5.8 Concluding Remarks

The aim of this chapter has been to examine the concept of the relation between cause and effect as it evolved in Leibniz's dynamics project. The key transformation was the move from *potentia* to *actio*. As *potentia*, *vis* was an intensive magnitude that causes motion through the exhaustion of intensity in extended motion. While the concept of *vis* qua *potentia* was never rejected, another notion of *vis* was foregrounded in the maturation of the dynamics project. This is the notion of *vis* qua action. Throughout this chapter we have argued for the difference between these two conceptions. *Vis* qua *potentia* required a theory of future effect and can be only treated as a final cause. *Vis* qua *actio* is temporally immediate and satisfies the notion of both final cause and formal cause. *Vis* qua *actio* also delivers on the idea of an immanent, formal cause of motion within a body or a system of bodies that goes beyond the notion of *vis* qua *potentia* where the intrinsic and external causation was undecidable from the perspective of mere conservation.

From these arguments, I have proposed a theory of structural causation as the best interpretation of the intention guiding the development of the dynamics project and its eventual conclusions. Structural causality in the dynamics requires that we grasp the structural property of the conservation of mv^2 as insufficient to constitute a theory of causation. It is only through the inherence of cause, the action of a body or system of bodies, that we can finally determine the rational core of the dynamics. The invariant and immanent causation of physical phenomena requires that the world is structurally produced synchronically by the immanent *actio* of bodies in a physical system.

But what is the subject of this action? What is the status of the body or bodies underlying the immanence of *actio*? In anticipation of the next chapter, let us point to an unresolved problem in this chapter. Given that we have accepted the interpretation of structural causation where bodies are immanently causal due to *actio*, how is this immanence indexed? If we follow the maxim "*actiones sunt [autem] suppositorum*," then actions and properties are only determinate when related to their *suppositum* or substrate, their substance.[6] In this, the idea of the immanence of *actio* still relies on an *ad hoc* idea of an individuated empirical body. This idea of the *actio of* a body has been operative in the many examples above. My argument here relies on the idea that the immanent cause and effect relation is indifferent to physical systems of one, two, or indeed *all* bodies. The subject of action, insofar as its causation is structural, cannot be localized spatially as a particular body. The universality of *actio* and the universality of the conservation of mv^2 means that a physical system of one or many bodies can serve as the subject of this immanent causation. The subject of action is a system of bodies ranging from the smallest parts of the created world to the greatest (just smaller than the totality).

From a physical perspective, we have already provided an answer that satisfies the subjective perspective of a single body. Given the equivalence of hypotheses and the invariance of *vis* qua *actio*, we can determine the coherence of physical phenomena from the perspective of the relative effects experienced by a body in the created universe chosen in an intuitive way. This does not, however, address how the immanence of *vis* can be individuated. According to a metaphysics that takes spatially localized substantial bodies, *vis* belongs to substances acting in space (and time). Here, the immanent *actio* of a body belongs to this cat or that clock. But if the doc-

---

[6] Leibniz provides a clear exposition of this position in article viii of the *Discours* (GP IV 432–433; AG 40–41). The axiom comes directly from Thomas Aquinas, ST II-II 58, Art. 2.

trine of *vis* is to be truly structural in causation, its immanence should be generic. That is to say, the structure of causation should be invariant across subjects. Although this point is very counter-intuitive, we can say that the structural causation of the dynamics is indifferent to the spatial localization or individuation of the subject of *actio*. As such, the ontology of *vis* is adaptable to Leibniz's shift from a metaphysics of spatially located bodies to that of a non-spatial monadic metaphysics. It is to this question of the subject of action that we shall turn to in the next chapter.

Chapter 6
Vis viva in a Monadic World

Abstract The sixth and final chapter of this book attempts use the perspective developed in the previous chapters to resolve a seeming contradiction between the doctrine of inherent substantial forces developed in Leibniz's dynamics and the late doctrine of the autarchy of monads, a mainstay of his late period (post-1695) metaphysics. The immediate problem is how a dynamical theory of corporeal motion can be coherent with the seeming reductive kinematic nature of physical reality implied by monadic perception. This problem is treated first by looking at Leibniz's theory of space in the analysis situs project and using its insights to understand the nature of relative motion in his correspondences with Clarke. This investigation sheds light on the concept of physical causality in the dynamics that treats the effects of physical phenomenon as a group of internally related empirical variations. Solidifying this concept of structural or dynamical causality by contextualizing it through the methodology of symmetry and invariance that Leibniz inherited from his mentor Huygens, this notion of causality can be shown to be consistent with the autarchy of monads.

6.1 Introduction

In the previous chapters, we have examined the principles of the dynamics and how they constituted the dynamics project as a whole. In the final chapter of this book, we will examine an instance of the relationship between the dynamics and the metaphysics developed in the same period. We examine the adaptability of the dynamics to Leibniz's late monadological metaphysics. The aim here is to grasp how the dynamics converges with Leibniz's late metaphysical system of monads and reconsider the relation between dynamical causality and its fundamental ontology. This is not an obvious task since the origin of the dynamics in the late 1670s was formed in a moment when Leibniz held a metaphysics of corporeal substances, a hylomorphic union between substantial form and matter. As the dynamics project progressed, the years of the late 1690s saw the rise of a different metaphysical system, one based on a metaphysics of monads, a monist ontology of simple indivisible substances.

The conflict between the metaphysics of the late 1670s and the late 1690s is namely that, in the world of autarkic monads, where there is no monadic (substantial) interaction, physical reality would appear to be reductively kinematic, and the "subjective" phenomenal representations of corporeal motion would be separated

T. Tho, *Vis Vim Vi: Declinations of Force in Leibniz's Dynamics*, Studies in History and Philosophy of Science 46, DOI 10.1007/978-3-319-59055-4_6

from its purported "foundation" in the realm of dynamical forces. On the other hand, the succession of these perceptions within the monad is entirely immanently caused by the appetition of monads driving towards a universal harmony. Now, the importance of the kinematic-dynamic, if one can forgive some anachronism, or effect-cause, distinction was central to Leibniz's dynamics project. As we have argued, the dynamics project itself began in the late 1670s as a reform of Cartesian mechanics concerning the laws of motion and collision. In this development, Leibniz was eventually able to put forth a foundational and long-held critique of the kinematic conception of physical reality (rightly or wrongly attributed to Descartes) reducible to *res extensa* (or size, shape, and motion). As such, Leibniz's development of the concept of *vis* aimed at providing a cause for motion that would not simply be reduced to phenomena like size, shape, and motion. This causal entity, *vis*, would then serve as the foundation and fundamental object of the dynamics as "*nova scientia*" (A III 4, 483–486). Hence, for the coherence of the dynamics, a merely kinematic account of motion was to be avoided at all costs. Locomotion is relational, indeterminate, and continuous. It is thus in need of a grounding in something that goes beyond size, shape, and motion. Yet the merely mechanical-geometrical view of motion appears to be exactly what the world of autarkic monads implies: a world of individuated monads without interaction that perceives the physical world as a series of successive images. These internal perceptions (harmonized with the internal perceptions of other monads) would seem unmoored from an objective or substantial counterpart and organized only through the immanent harmony of substance itself. Hence despite Leibniz's explicit attempt to place his dynamics, a science of *vires*, within the monadic universe, by identifying "primitive motive force" (*vim motricem primitivam*) with the ἐντελέχεια (*entelechia*) constituting the form of substances (substantial form), the connection between the cause of motion and the appetition of monads would appear to be only a metaphysical one (GP IV 511; AG 162). This approach leaves little room for the dynamical and physical account of locomotion within this monadic system.[1]

In view of this problem, there is one possible interpretation that first needs to be set aside. Many recent commentaries on the metaphysics of Leibniz's so-called "middle years" (circa 1678–1695) see Leibniz as embracing some form of physicalism via the rehabilitation of quasi-Aristotelian-Scholastic substantial forms.[2] In this view, a matter-form hylomorphism constitutes a real corporeal substance, and Leibniz's dynamics would directly apply to physical phenomena. In this interpretation, the hylomorphic unity of form and matter would neatly divide the active (*vis*) and passive (resisting inertial matter) of corporeal substance respectively. In turn, physical reality would be constituted by the interactions between these corporeal substances localized in space and time. It is precisely this interpretation, here being set aside, that makes it difficult to see the dynamics as coherent with the later monadic metaphysics. The exact nature of Leibniz's metaphysical "turn" from privileging corporeal substance (via the doctrine of substantial form) to non-corpo-

---

[1] See Leibniz, *De Ipsa Natura* (GP IV 504–516; AG 155–169).
[2] See Garber 1985, 27–130 and Fichant 2004.

real autarkic monads is disputed, and I will not intervene in this metaphysical debate here. It seems, however, that the metaphysics of the dynamics is more suited to the earlier (middle years) metaphysics of substantial forms than to the later monadic metaphysics. According to the interpretation of a hard shift from corporeal substances to ideal and simple monads, it seems that Leibniz had to make the dynamics fit into the new metaphysical picture by awkward *bricolage*. The *vires* that once animated real physical entities would, in the world of monads, *merely* regulate the (kinematic) appearance of physical entities.

This interpretation of a shift between different metaphysical systems between the "middle years" and the late (starting around 1695) is controversial. Many interpreters see the focus on corporeal substances in the 1680s as merely an issue of emphasis. Still others see Leibniz's emphasis on the organic body in the later period as the lens through which to harmonize Leibniz's many overlapping metaphysical systems.[3] We cannot address this enormous problem of interpretation here. It is mentioned only to highlight the obvious difficulty of articulating the foundations of physics where the entities of Leibniz's fundamental reality are non-physical and non-interactive. With this difficulty in mind, I set aside these metaphysical problems in order to examine the dynamics as a separate project that is adaptable to different metaphysical systems. The aim of this chapter is to argue that, not only do *vires* inhabit the monadic world comfortably, but also that we might see the metaphysical implications of the dynamics as more satisfactorily fulfilled by monadic metaphysics. The internal metaphysical foundations of the dynamics is adaptable to different metaphysical commitments with portions that were exploited to make different foundational claims about substance at different times. Hence although the dynamics seem to be more suited to a world of corporeal substances, the aim here is to illustrate that, seen through the lens of dynamical causation, it is at worst orthogonal to this disputed metaphysical shift in doctrine or, at best, an important contributing factor for the fullness of the monadic world.

The solution proposed, in brief, is to show that the dynamics itself relies on a deep and fundamental distinction between cause and effect *qua vires* and extended motion. This requires that we reaffirm the concept of causality that we have been developing throughout this book. Dynamical causality, as we have argued, cannot be understood as the local empirical or mechanical-contact relation between moving bodies but should rather be understood as the relation between a non-phenomenal realm (of *vires*) and a spatial-temporal *system* of phenomena. Hence, dynamical causes and phenomenal effects are ontologically distinct. As we have developed this concept over the last few chapters, this form of causality is understood as "structural" simply because cause and effect operate across different levels of reality. In this model of physical causation, *vires* produce locomotive effects through a distribution of quantities governed by architectonic laws. Most central among them is the invariance of the measure of *vis* (*vis viva*) mv^2 the product of mass (m) and the square of velocity (v^2). Dynamical causality here operates through the invariance that governs internally related systems of bodies. Hence, the non-spatial and structural reality of *vires* is

---

[3] See Duchesneau 2010.

both consonant with the autarky of monads and causally responsible for locomotive phenomena. *Vires* thus constitute the "objective" counterpart of "subjective" monadic perceptions precisely because they determine and ground the merely apparent, though intelligible, mechanical interactions between bodies in phenomena. Without impugning the *intelligibility* of mechanical interaction, the true cause of motion is *vis* and not mechanical *power* [*potentia*]. Hence the dynamics, which predates Leibniz's adoption of the monadological metaphysics, can be understood as carrying through and contributing to the later monadological picture.

I shall make this argument along the following lines. In the second section, I will provide some contextual clarification by looking at the difference between "relative space" and "relative motion" in the period of monadic metaphysics. By reinterpreting key problematic passages in Leibniz's dispute with the Newtonian Samuel Clarke through Leibniz's *analysis situs*, I argue that the objectivity of space, not unlike *vires*, is to be found in the structure of phenomena rather than the actual phenomena themselves. This is crucial for understanding how the non-local and non-spatial reality of *vires* causes and determines local and spatial phenomena. I will also make the case for the understanding of Leibniz's "relativism" as being more accurately understood as a relationalism of motion rather than a relativism of space.

In the third section, since this relationalism of motion (rather than space) is the key for grasping the bridge that Leibniz built between empirical phenomena and its causes, I shall reiterate some of the arguments already presented in this book for understanding the relationalism of motion and its relevance for the structural nature of dynamical causation. The relationalism of motion strictly means, following Leibniz's concept of the equivalence of hypotheses, that any locomotive phenomena can only be understood through the invariance of *vis viva* as the structure of locomotive events. By providing some contextualization of Leibniz's methodology in the dynamics, I will demonstrate how Leibniz borrowed from his mentor Huygens in order to employ the concepts of invariance and symmetry for establishing the key relation between *vires* and locomotive phenomena. This methodology, I argue, provides the framework of the central concept of structural causality at work in the dynamics.

In the fourth section, I use the analyses developed in the previous sections in order to clarify Leibniz's treatment of complex empirical phenomena, such as centrifugal and celestial rotational motion. Here, despite the large number of concrete mechanical questions that Leibniz took up between his Paris period (1672–1676) and the late 1690s, we should be careful in identifying the concept of the *cause* of motion that formed the core of the dynamics and guided its development. The dynamics explains different mechanical phenomena and the extension of these explanations to machines and astronomical bodies. However, confusing principle and application in the dynamics muddies the understanding of the relation between, on the one hand, Leibniz's theory of dynamic causality and, on the other hand, Leibniz's more general concern to provide intelligible mechanical explanations. This clarification will help distinguish Leibniz's treatment of dynamic causation and empirical phenomena and clarify the "structural" nature of the causation at work in the dynamics.

In the fifth section, I argue for a synthesis of these elements of the dynamics by showing their harmony with the monadic vision of substances. The aim here is to show that the internal perceptions of autarkic monads are not only externally harmonized through the doctrine of God's creation of the best of all possible worlds, but also that the inherent activity of the monad is responsible for the nature and structure of physical reality. This immanent affectation of the monad, understood in various presentations as *entelechy*, action, or appetition, is thus consonant with the central aims of the dynamics: a systematic and foundational account of physical reality that does not reduce to kinematics.

6.2 Situation, Space, and Motion

Before entering into the dynamics, we shall begin with the period of the late monadic metaphysics through an examination of Leibniz's development of the concept of space. It is important to begin with this discussion of the Leibnizian concept of space because his dynamics has too often been treated solely through the lens of its alluring (but illusory) offer of an alternative to Newtonian classical mechanics and its accompanying notion of absolute space.[4] Part of the motivation here is to develop a new vantage point to distinguish problems of space and problems of motion in Leibniz and also to move the discussion away from unwarranted opposition between Newton and Leibniz that have too often colored Leibniz interpretation. As we shall see, it is true that Leibniz, Newton, and the Newtonians differ on the central problem of the ontological priority of space over extended bodies in the account of physical motion. Nonetheless, they do not differ on the problem of the structure of space or the logical priority of this structure over spatial things. As I will argue in this section, all parties generally agree on the logical priority of space over extended phenomena. Consequently, they do not differ on the geometrical properties of motion, a question that is orthogonal to the ontological priority of space or extended bodies. Clarifying this point will help distinguish the relativity of space from the relationalism of motion and thus aid in the understanding of dynamical causation.

Leibniz had already stated at the very beginning of his Paris period, in the 1672 *De Minimo et maximo*, "There is no space without body, and no body without motion" (A VI 3, 99; LC 15). This neat expression does sum up the *relative* and *ontological* dependence of space on bodies and, in turn, bodies on motion and this relation remains unchanged in Leibniz's work after this point despite taking up different interpretations. Nonetheless, the of motion refers to something entirely different from that of the relativity of space. On the one hand, a relativity of space, for the post-Einsteinian age, refers to the variation of spatial structure (curvature), but for Leibniz, Clarke, and many contemporaries, as we shall examine below, it refers to the ontological and logical conditions for the existence of spatial things. On the other hand, the relationalism of motion refers, in a strict sense, to Galilean relativity or invariance, formulated by Leibniz as the principle of the equivalence of hypotheses:

---

[4] See Reichenbach 1928.

[T]o whichever body we might in the end truly ascribe motion or rest, the same outcome
would be found in the phenomena in question [...] the same outcome would be found in the
resulting phenomena, even as regards the action of bodies on one another. (GM VI 247; AG
131)

Hence, in order to understand this central pillar of Leibniz's dynamics, it is first
necessary to clarify the nature of relative space and the role that it leaves for an
entirely different kind of "relativism."

Although many commentators approach Leibniz's philosophy of space almost
exclusively from the writings surrounding his correspondences with Clarke, an
independent line of research on space was part of Leibniz's work long before the
monadic period. Leibniz had adopted the equivalence of hypotheses as early as the
1676 essay *Principia mechanica*, where Leibniz stated boldly that "motion and rest
taken absolutely are empty names, and whatever is real in them consists in respective change alone" (A VI 3, 110; Leibniz 2013b, 115). Along these lines, the quali-
fication of physical space as only real through the relation between bodies carried
through in treatises in the 1677 "Spatium et motus revera relationes" (A VI, 4,
1968–1970; LC 225–227). Nonetheless, relative motion is one thing, but space is
quite another. Here we must turn to the *geometrical* problem of space that was
treated by Leibniz in an extensive project which is generally called the "*analysis
situs*" (after his 1693 text made famous by Poincaré's use of the same title for his
1895 founding paper of algebraic topology).

In my brief look at the *analysis situs* project, I owe much to Vincenzo De Risi's
elegant and profound study of the subject in his 2007 *Geometry and Monadology*.
The *analysis situs* is to be historically contextualized as a project of the rigorization
of geometry and historically classified alongside similar works by near contemporaries Roberval, Barrow, and Wallis in the age following Descartes' ambitious reorganization of geometry through algebra.[5] These projects all sought to make geometry
more rigorous through the demonstration of foundational concepts and methods,
such as the semantic determination of the point, line, and circle. The focus of such
projects concerned, for example, the consistency of the Euclidean definition of the
line spread between the definitions of Book I of the *Elements*. The reception of
Euclid had historically generated a number of different meanings for these fundamental figures by commentators and transmitters like Proclus and Clavius.[6] This
historical situation produced a series of different kinds of problems to which Leibniz
answered with a geometry of pure "situation," that is, "position" or *situs*.

The original terminology of *situs* already played a role in Leibniz's early thinking
in the 1666 *De Arte Combinatoria*. At age 20, Leibniz did not yet have the explicit
geometrical use of the term *situs*, and it was understood logically as reciprocal positions of symbolic permutations relevant to the (logical) art of combination. The evolution of *situs* was eventually applied to geometry and the abstract theory of space
starting in the 1671 *Elementa de mente et corpore* and, through the Parisian period
(1672–1676), would mature into the attempt to reconstruct (Euclidian) geometry
itself. The real maturation of the project came in the post-Parisian period when the

---

[5] See Roberval 1996, Barrow 1860 and Wallis 1695, 11–228.
[6] See Proclus 1970 and Clavius 1611–1612, 13–638.

analysis situs project came into full swing with Leibniz's composition of major texts aimed at reconstructing the fundaments of Euclidian geometry. De Risi identifies 1679 as the *annus mirabilis* of the *analysis situs* project (De Risi 2007, 63).[7] Unfortunately, in Leibniz's correspondence with his mentor Huygens late in that same year (1679), Huygens had so discouraged Leibniz's project that a decade gap existed before Leibniz would again feel confident enough to disseminate his work on the project despite occasional work on it. The embattled Leibniz would eventually regain his compass in the 1690s and again begin to correspond on the topic. This "late" work constitutes his most significant publications on the topic, including the major texts "*De Analysi Situs*" (1693), "*Specimen Geometriae luciferae*" (1695), and "*In Euclidis πρῶτα*" (1712). There is no doubt that this significant and sustained project formed the background of Leibniz's use of the terminology of *situs* in correspondence with Clarke. Indeed, the whole of the spatial theory in this period of Leibniz's thinking can also be closely associated with the massive undertaking of a philosophy of the foundations of geometrical knowledge and its theoretical entities.

Without going into the details of the *analysis situs* project, we note that, in the maturation of the project, Leibniz developed a foundational theory of geometry based on the minimalistic foundation of unextended points and their possible relations available from rigorous logical definition. This is the basis upon which the reconstruction of Euclid was attempted. As simply one important and characteristic example from this large body of work, in the "*In situ est extensum et extremum*" of 1695, we find that what is characteristic of this project was that the successive definitions for point, solidity, line, and the like were constructed from logical definitions of *extensum* and *extrema* (the former "bounded" extensions and the latter "boundaries") in order to build up the fundaments of geometry. Here, the point, the line, the surface, and the solid are successively defined with respect to their logical possession of bounded figures (*extensa*) and boundaries (*extrema*). A point, for example, is defined as a boundary (*extremum*) that does not itself possess a boundary (*extremum*) and a line is defined as that which does not possess a boundary (*extremum*) that is (itself) extension (*extensum*) (De Risi 2007, 207–208). Ironically, we find here a seventeenth-century analogue of a Hilbert rather than a Poincaré, as the project of rigorization was pursued with logical reduction in mind. De Risi notes that, despite Leibniz's dissatisfaction with this text of 1695, he continually returned to this form of logical-constructive presentation (De Risi 2007, 207).

Leibniz's attempt to standardize Euclidean geometry led to an issue more salient to our concerns. That is, once we take points and their relative positions (*situs*) themselves to be primary, the ambient space upon which these points lie becomes foregrounded. Certainly these ideas may suggest a connection first, to non-Euclidean geometry and, second, to algebraic topology. Of course, as De Risi points out, neither of these important developments in mathematics could directly be imputed to Leibniz's work (De Risi 2007, 114).[8] Nonetheless, Leibniz did widen the conceptual gate for considering geometrical structure solely from the starting place of the extension less point.

---

[7] See Leibniz 1995.

[8] See also De Risi 2015, 1–13.

Fig. 6.1 Leibniz's
definition of a plane using
two points in the *analysis
situs* project (This figure is
taken from De Risi 2007,
216)

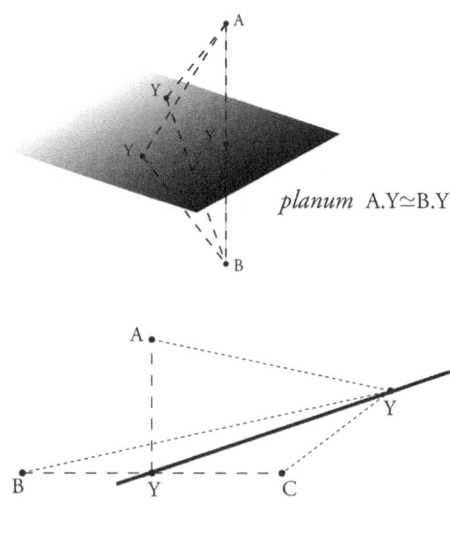

planum A.Y≃B.Y

Fig. 6.2 Leibniz's
definition of a line using
three points in the *analysis
situs* project (This figure is
taken from De Risi 2007,
217)

recta   A.Y≃B.Y≃C.Y

In order to give a small taste of the content of the *analysis situs* project for a philosophy of space, I draw attention to Leibniz's definition of the line and plane in the mature work of the *analysis situs* project such as the "*Rectam definio...*" of 1715 (De Risi 2007, 612–615). The argument can actually be found in earlier works dating back to 1679, but we make our sketch of the theory from the later work. The idea here is that the plane can be defined as the locus of (the set of) all points that bear the same situation with respect to *two* given points (De Risi 2007, 215). In turn, a line can be defined as the locus of (the set of) all points that bear the same situation with respect to *three* given points.

Given a figure with points A and B, there is a set of Y points determined by the congruence between A and Y, and B and Y (Fig. 6.1).

Hence, the plane determined by the points A and B is constituted by the set of points that satisfy A.Y≃B.Y. In turn the determination of the line will then be given by *three* points and the situations between of A, B and C where A.Y≃B.Y≃C.Y (Fig. 6.2).

The idea here is that the determinations of boundary (*extremum*) and extension (*extensum*) can be constructed with the careful handling of the *situs*. Leibniz thus envisions a theory of space that builds from logically dependent constructions from point to line, plane, and solid.

Despite the portentousness of Leibnizian , this project remained limited. Of course, despite Poincaré's reference to this body of work in homage to Leibniz, any idea that Leibniz had a direct continuity to such later developments would be mistaken. The issue is that, although Leibniz attempted to found geometry on the basis of unextended points, the relation between these points remained squarely metric through the notion of distance that constitutes the *situs* relation between points and the underlying methodology of extensional equality or identity remained tied to congruence. In De Risi's definition of Leibnizian space, he emphasizes that it is a "non-compact isotropic complete connected oriented three-dimensional Riemannian manifold without boundary and with zero Riemannian curvature" (De Risi 2007, 263–264). For our purposes, we can simply remark that the abstract space defined by Leibniz's *analysis situs* is Euclidean-3 space and its subspaces.

This brief look at *analysis situs* provides us with a baseline for our investigation here. Space, for Leibniz, is constituted by situations (mutual distances defined by A.Y≃B.Y, etc.), indexed by unextended points. The structure of this space is given by the relative distances, governed by congruence, between those points and other geometrical relations (like similarity) that exist between situations. In turn, higher dimensions (dimensions up to three) are given by those same relations through the principle of homogony. This concept determines figures of higher dimensionality through their determination by points. This concept of space then informs us of the crucial role, in his assertion to Clarke, played by the idea that space cannot be understood *as* a situation but also cannot be understood as *sui generis* (that is, without situations). Space is rather the *order of situations* rather than just *order* or *situations* (GP VII, 415; Leibniz and Clarke 2000, 60). In three dimensional terms then, the spatial world is the set of all the relations, or the order, between possible situations governed by Euclidian-3 space.

Following Leibniz's geometrical project, we know that his final theory of space, as we have attempted to summarize here, is the "set of all abstract situational relations" (De Risi 2007, 561). De Risi proposes a transcendentalist interpretation of Leibniz's metaphysics of space in order to make sense of the continuity of the *analysis situs* project and his various contemporaneous physical and metaphysical (monadological) theories. For De Risi, Leibniz's transcendentalism partly means that the abstractness of the order of situations is the ambient spatial structure of the internal representation of the relation between *situs* in any given perception. It is the condition of the possibility of any given spatial phenomenon. This absolute space is absolute because, for any given spatial representation, Euclidian geometry holds. Now, De Risi's "transcendental" terminology here may strike some as anachronistic (foretelling Kant's "critical turn"), so I will use the more neutral terminology employed by Arthur, which is taken from a Leibnizian manuscript from the late 1670s: *generic* space (*spatium generalis*) (A VI, 4, 1397; LC 241–243).[9] The point is the same: although Leibnizian motion is (concretely) relative, *generic* space is (abstractly) absolute.

An important point follows from this look at the *analysis situs*. The "philosophy of space" developed by Leibniz over many long years was meant to provide a foundational theory of geometry. As such, the question of space remained limited to the domain of the imaginary realm of spatial reality. Since Leibniz understands the physical world to be expressed spatially by the relations between extended objects, physical phenomena, insofar as they are phenomena, are obviously subject to the abstract determination of spatial structure. Insofar as Euclidian-3 space is necessary, such a space allows us to bring Leibniz's geometrical studies to bear on the necessary conditions for the possibility of locomotive phenomena. Leibniz insists on this point in the 1703–1705 *Nouveaux essais sur l'entendement humain*,

---

[9] See Arthur 2013, 506–507.

> [T]here is no need to postulate two extensions, one abstract (for space) and the other con-
> crete (for body). For the concrete one is as it is only by virtue of the abstract one: just as
> bodies pass from one position in space to another, i.e. change how they are ordered in rela-
> tion to one another, so things pass also from one position to another within an ordering or
> enumeration. (A VI 6, 127; GP V 115; Leibniz 1981, 127)

We examined this mature view of space in Chap. 4 when we considered the status
of continuity in the dynamics project. Here again we emphasize that the spatial relations in physical phenomena *simply are* those relations determined by this necessary
(logically prior) and generic space. The structure of locomotive phenomena is thus
logically conditioned by the structure of this generic space.

The analysis above allows us to gain some precision in what we mean when we
speak of Leibnizian relative *motion*. But first let us grasp the specificity of Leibniz's
aims for the concept of relative *space* through the evolution of his argument with
Clarke. At the start of the correspondence, the only problem that Leibniz posed to
Clarke was the problem of space as *sensorium dei*. It is after Leibniz's third letter
that we concretely move away from the theological questioning to a more scientific
one. Clarke's response provided the scenario where the whole creation was moved
in a straight line akin to the shift of (inertial) reference frame. For Leibniz, the spatial shift of the totality of the world fails to be spatial since such this shift would
leave relative *situs* unchanged. Leibniz puts forth this position in the fourth letter by
arguing that, "For two states indiscernible from each other are the same state, and
consequently, it is a change without any change" (GP VII 373; Leibniz and Clarke
2000, 23). This response of course also covers Clarke's earlier challenge concerning
the mirror-like inversion (isometry) of the universe east to west.

In this context, invoking the principle of the identity of indiscernibles, Leibniz
argues that, since the relative positions and relative states of the bodies under consideration under these different scenarios cannot be distinguished, the entire configuration of *situs* is indistinguishable between the two scenarios. However, this
consideration says something more than the mere inversion of situations "east and
west" or the movement of inertial frames which preserve situational relations. It
implies motion, and Leibniz's invocation of indiscernibility with respect to inertial
frames invokes the dynamics and its principle of the equivalence of hypotheses
previously defined. Hence, we should underline that it is motion that stands as relative in this statement rather than space itself. If space were determined by the set of
possible relative distances between bodies, these *situs* would themselves be invariant. Whether a set of situations inverted "east and west" or shifted according to
inertial frame, relative distances remain mutually identical (indiscernible). However
in the shift of inertial frame, the true problem of *relationalism* can only arise through
the respective distribution of rest and motion for a physical event more than mere
static situational relations. Hence, Leibnizian relative motion must be understood as
a united system of variations of rest and motion according to the principle of the
equivalence of hypotheses. With the relative *variations* of motion and rest, relative
space is not salient to the discussion.

Now, of course, when we speak of the "relativity" of *space* in Leibniz, we speak of the *ontological* dependence of space on bodies. To this Leibniz states unequivocally to Clarke that,

> It is true, it [space] does not depend on such or such a situation of bodies, but is that order which renders bodies capable of being situated, and by which they have a situation among themselves when they exist together. But if there were no creatures, space and time would be only in the ideas of God. (GP VII 376; Leibniz and Clarke 2000, 27)

There is perhaps no clearer statement of Leibniz's mature ontology of space. Without the creation of localized bodies, space would not exist. At the same time, this generic space is the order of situations that renders bodies capable of being situated. Space *qua* order of (possible) situations is logically prior to the localization of bodies. This means space is absolute in a generic and abstract sense despite being ontologically dependent on existing bodies. Leibniz continues in this mode of response until the end of the correspondences (Leibniz's death in 1716), and Clarke continues to assert the problem of the spatial shift of the totality of the created world.

In their correspondences, Leibniz does not answer Clarke with anything more than a hint at the enormous work that had led up to his mature position concerning *situs*, generic-absolute space, and the dynamics. Although relative motion was certainly one of the themes of the debate, the problem was overlooked in order to focus on the ontological problem of the reducibility of space to situations. The debate with Clarke and the meaning of "relationism" and "relativity" here is only that if there are no bodies, then there are no actual situations, and hence there is no actual space. This logically entails that actual corporeal bodies in space have a determinate order with respect to relations constituted by their mutual distances defined in generic space. Nonetheless, this determination of the generic structure of space as the order of situations is absolute and the problem of the position of actual bodies in motion is orthogonal to this determination.

6.3   The Determination of Relative Motion: Invariance and Symmetry

The examination of Leibnizian space above addresses the difference between relative space and relative motion. The "relativity" of space refers to the ontological dependence of space on the existence of bodies. Yet what is the nature of this "relationalism" of motion here? We shall move towards answering this question with a look at the dynamics. In this I shall provide only a partial account of Leibniz's dynamical causation that has been treated at greater length in previous chapters.

In the dynamics, Leibniz systematically used the principle of the equivalence of hypotheses to define relative motion. We have already treated this principle in previous chapters. This principle states that for any system of *n* bodies, we can individuate the same physical event regardless of which of the *n* bodies in the system we take to be at rest or in motion. This is the kind of generalization of Galilean relativity already developed by Leibniz's mentor, Huygens, who developed the concrete theory as early as the 1650s and laid out the principles in his 1669 publications in the *Journal des Sçavans* and *Philosophical Transactions*, which were summaries of his then unpublished treatise *De Motu Corporum ex Percussione* (Huygens 1929, 30–91; 1977, 574–597).

As we have discussed at length in Chap. 4, this principle of the equivalence of hypotheses states that any hypothesis concerning the relative motion and rest of bodies in a physical system is equivalent under the condition of the symmetries of collision and conservation of *vis*. Hence motion is in the ultimate sense *relative* given the "equal" truth of various hypotheses. As such we must understand physical events as a series of variations according to which a system of n bodies is given relative states of rest and motion. Motion is thus relative insofar as it can be represented in a series of internally related distribution of measurements governed by conservation. This implies that a system of a single body can be taken either at rest or in any degree of motion, clearly different phenomenal effects, while implying the same cause. In turn, motion within a system of more than one body is fully determined across a variation of different distributions of velocities among the bodies governed by the invariant mv^2. Here the quantity of *vis viva* mv^2 is important for the determination *via invariance* of these internally related groups of variations.

What is crucial here is only that *vis viva* constitutes the invariant for this group of variations, regardless of the magnitude of the conserved quantity. In the scope of possible worlds, mv^2 need not be the measure of *vis viva*. In a purely conceptual sense, the causal nature of Leibnizian *vires* does not require much else than a group of variations and its invariance. But the rigor of the concept required that Leibniz provide a concrete bridge between the concept of *vis qua* cause and the quantitative structure of actual motion. Leibniz inherited much of this from his mentor, Huygens, whose demonstrations in various treatises, closely read by Leibniz, served as the basis for the development of the dynamics. Central to this inheritance was the Huygensian methodology of the center of mass frame.

As we have argued in previous chapters, the center of mass frame or center of gravity was the methodology that Leibniz relied on to build a bridge between non- or infra-phenomenal cause and phenomenal effect. The underlying assumption that makes this methodology a powerful one is its extension of Galilean relativity. Galilean relativity allows us to determine invariances between motions in different relatively determined inertial frames. The methodology of the center of mass frame in turn allows us to determine these same invariances across arbitrary inertial frames. Of course, for the modern reader of classical mechanics, these issues tend to be reducible to the single idea of the center of gravity or center of mass frame. Nonetheless, the additional step made by Huygens was important to Leibniz, who owes this central dynamical tool to this Huygensian methodology.

Huygens' methodology consists in showing that we can determine the relative initial, average, and terminal velocities of a system of bodies in collision in terms of the symmetry of the center of mass frame. This in turn allows us to determine absolute differences of *potentia* or *vis* between different systems. Huygens' crucial demonstration comes at the end of *De Motu Corporum ex Percussione*, following the main demonstration of relative velocity reversal in elastic collisions within Galilean relativity. We have examined this argument in previous chapters. What is important for us here is to recall that, beyond its elegant symmetry, relative motion allows us to determine the conservation of mv^2 and mv. This methodology also provides for the demonstration of the universalization of the concept of Galilean relativity governed by this symmetry. That is, Galilean relativity provides the grounds for measuring absolute magnitudes of *vis* beyond inertial motion and collision. This extension of the Galilean principle is the crucial concept inherited by Leibniz through the Huygensian method. Of course it also indicates the statical limits of Leibniz's conception, which were precisely the same as Huygens'. Hence although Leibniz understood the conservation of what we now call energy-work, he only intuited, as I discuss below, the temporal evolution of the transformation of the conserved quantity, mv^2, in his laws of motion.

This statical limitation of Leibniz's concept of *vis* is important because his theory of motion (following Huygens) stops short of what we now call "dynamics." Despite Leibniz's various attempts to provide a time-dependent account of the translation of *vis* to motion in his later work, the core of his theory remained tethered to the static methodology of symmetry. That is, we find a theory essentially tied to initial, final, and average velocity. This is the constant theme of Leibniz's major statements of the dynamics. Nonetheless, an argument concerning the evolution of motion in time is available in the late document of the dynamics, *Essay de dynamique*.

As we have examined in previous chapters, the *Essay de dynamique* (*circa* 1699–1701) is the last comprehensive contribution to Leibniz's dynamics project (GM VI 215–231). It followed the turn in the mode of presentation brought forth after the *Dynamica* where the term "*actio*" was introduced to provide a new presentation of *vis* in terms of its embodiment in the immanent evolution of corporeal motion in space *and* time. The innovation of this presentation of *vis* through the concept of *actio* is the inclusion of a time factor in his demonstrations. This presentation remained tied to the statical methodology borrowed from Huygens but sought to reach beyond it. Leibniz's strategy here is to divide the measurement of *vis* up into two factors, first a factor of "formal effect" measured by the product of mass and distance ($m \cdot s$) and a factor of velocity (rate of distance over time s/t) taken at some time t_n (GM VI 222). *Actio* would then be the product of formal effect and velocity ($a = m \cdot s \cdot (s/t)$) taken at different times t_n. Since velocity is distance s over time t, *actio* over time can also be understood as mv^2. Hence for the calculation for an invariant in a system of bodies in motion, the calculation of mv^2 at time t_n is equivalent to the calculation of *actio* at time t_n.

With this presentation of the measurement of *vis* through *actio*, Leibniz's aim here is to model a system of bodies in order to demonstrate the invariance of *actio* or *vis* at each time t_n as the system evolves. By assigning different speeds and masses to each body and in tracking their speeds at time t_n, Leibniz seeks to make his case for a conservation of *actio* or *actio motrice* (GM VI 220). The goal of an invariant calculated in this way allows us to look at the quantity mv^2 differently. We know that since the respective speeds of each body in a system are proportional to the distance traveled by each body, their speed is also linearly proportional to the "formal effect" ($m \cdot s$) of each body considered. As such, the quantity mv^2 is thus proportional to the quantity of motive *actio* of each body in the system at each time t_n.

This development shows Leibniz intending to present a concept of *vis* that goes beyond the statical methodology inherited from Huygens towards a theoretical understanding of the transformation of *vis qua* energy-work in time. However, it would be too generous to Leibniz to grant him a systematic understanding of the temporal evolution of the differential of work and the quantity of action, as we would find in Maupertuis and Lagrange, in this late writing of the dynamics. Nonetheless, it is clear that Leibniz had moved beyond the concept of *vis* as pertaining to mere intensive mechanical power (*potentia*). This transformation can be philologically noted in the transformation of terminology where the privileging of power or *potentia* is replaced by action (or *actio*) in the crucial period of late 1689.

The terminological move from *potentia* to *actio* has been noted by some careful commentators and in previous chapters.[10] Much more of scientific and metaphysical import could be said about this transition, although here I only want to emphasize this maturation of the concept of *vis* from *potentia* to *actio* for the purposes of underlying my basic contention that the causal nature of *vis* concerns the symmetrical transformation of quantities of motion under Galilean relativity rather than the mechanical intensities (*potentia*) within physical systems that produce extensional effects. In this view, the pendulum swing, often used in Leibniz's dynamical demonstrations, that translates the intensive quantity of motion (velocity) to the extensive quantity of height cannot be merely interpreted as the translation of a power (*potentia*), through its exhaustion, to extension. Rather, these statical quantities (initial, final, maximum, and average velocities) represent groups of related effects determined through the quantity of *actio* that corresponds to an integral physical system of one or many bodies in temporal evolution. As we noted, despite Leibniz's methodological limitations, the shift in terminology from *potentia* to *actio* is captured by this intention to elaborate a theory of the structural concept of causation.

6.4   Physical Phenomena and Their Causes

As we have emphasized in a number of places in previous chapters, it is very easy to misunderstand Leibnizian *vis*. The unstable definition of the term "*force*," in the seventeenth and early eighteenth centuries, hovering between Cartesian quantity of motion (mv), Leibnizian *vis viva* (mv^2), and Newtonian change of momentum ($mvdv/ds$), produced a theater of disputes that might be understood, as D'Alembert put it, a "dispute of words."[11] I take this dispute to be meaningful despite the equivocation over the terms "*force*," "*vis*," or "*Kraft*." This dispute is meaningful precisely because it allows us to grasp the specific notion of causality underlying these different attempts to provide systematic laws of motion. If we fix our terms according to (currently understood) classical mechanics, what we call (Newtonian) force is the cause of change of momentum, and this operational concept clearly does *not* correspond to

---

[10] See Fichant 1995 and Robinet 1984.

[11] See Hankins 1990, 207.

Leibnizian *vis*. However, with the maturation of the dynamics project, Leibniz was able to define the primary object of the *nova scientia* dynamics, *vis*, as the cause of motion (A III 4, 483–486). Hence, insofar as the idea of *vis qua* cause of motion had developed through a structural rather than operational notion of cause, the scientific occupation of the dynamics concerns the translation of *vis* into groups of corresponding extended physical phenomena rather than the geometrical description of the path of a unique (abstract) body in motion. The significance of this development and its difference from classical mechanics can be seen in two important and related components of Leibniz's work on motion throughout this period. The first concerns rotational motion, and the second concerns celestial vortex motion.

Since my task here is disambiguation, I cannot enter into all the details of Leibniz's treatment of rotational motion (discussed in Chap. 3), the calculation of centrifugal-centripetal forces, and the problem of celestial vortex motion. Here I only wish to establish the case that, although these issues were important for Leibniz, they were nonetheless *second-order* problems for the dynamics dealing with the nature of phenomena rather than causes. The aim here is to clarify the priority that the systematic theory of *vires* had over *kinematics*, or the account of the geometry of moving bodies.

Leibniz had held a metaphysically-informed position of relational motion long before encountering the writing of Newton and the Newtonians. The specific dating of Leibniz's reading of Newton's *Principia* is ambiguous, but Leibniz's relationalism of motion dates at least to 1677 (A VI, 4, 1968–1971; LC 224). Nonetheless, upon his realization that he needed to provide a robust response to the Newtonians, it was clear to him that it was the problem of rotational motion that needed most attention. With the motivation of finding strong anti-Newtonian allies on the continent, Leibniz resumed his correspondence with Huygens in January 1688 after a nearly decade-long pause (A III, 4, 368–371). In this recommencement, Leibniz was certainly aware of the strength of Newton's argument for the absoluteness of motion in his two examples of rotating globes and the spinning bucket. Rotational motion understood as acceleration certainly constituted a strong case for absolute motion insofar as acceleration was independent of Galilean invariance. Within this context, both Leibniz and Huygens agreed that the appropriate riposte was to develop an account of rotational motion that would decompose it into (infinitesimally) smaller rectilinear motions, tangents of the curve, and hence reduce the phenomenon of acceleration into "smaller" phenomena of non-accelerative linear motions. These decomposed rectilinear motions would be subject to Galilean invariance even if their compositions express acceleration. Huygens and Leibniz went about this task in different ways but the crucial point here is that Leibniz saw the solution to the problem in a geometrical rather than dynamical way. In other words, Leibniz met the Newtonian challenge of rotational motion without modifying his theory of dynamic causality.

The problem of rotational motion was, of course, also crucial for explaining celestial orbits, and here we see the extension of such a problem of rotational motion to larger implications of universal gravitational attraction. Again, there are complexities in Leibniz's account of elliptical celestial orbit and the inverse square law on which I will not expound. It suffices to note that Leibniz accurately reproduced the Keplerian area law in his treatment of harmonic circulation by the decomposition of rotation into rectilinear motions describing trans-radial velocity.[12] This is a clear case of the geometrical *description* of celestial orbits, and Leibniz instrumentalizes the geometrical account of celestial orbits in order to immunize his causal theory of *vires* from Newtonian gravitation. Here, rotational motion is reduced to momentaneous linear motions formed by tangents on the elliptical curve. The reason for the elliptical shape of celestial motion is, in turn, produced by the subtle vortex-motion of the plenum against which these inertial rectilinear motions deflect.

To press further on the important difference between Leibniz's empirical and dynamical aims, I take a brief look at the correspondence between Bernoulli and Leibniz in 1695 over the relation between plenum mechanics and Leibniz's dynamics. In the correspondence of 8/18 June 1695, Bernoulli remarks that he had read the *Specimen Dynamicum*, published in April of the same year in *Acta Eruditorum*. He poses a challenge, stating that if the plenum theory is part and parcel of the dynamics, the Galilean law of falling bodies or, more specifically, the "laws of gravity" between celestial bodies would be contingent on "the motion of ethereal matter" (A III 6, 398–411). Echoing Newton's *Principia Mathematica*, Bernoulli implied that, since Leibniz's demonstration of the conservation of mv^2 in the *Specimen Dynamicum* (and indeed for any empirical demonstration) cannot discount the motion of the plenum, this calculation for *vis* and, more specifically, the quadratic proportion height and velocity used in the demonstration in *Specimen Dynamicum* would be accidental since the shape of vortical motion is itself accidental (A III 6, 408–410).[13] Commentators like Garber have cited this challenge by Bernoulli in framing the kind of difficulty that we have been treating here: the ambiguity between the core theory of the dynamics and its application in treating difficult cases of empirical motion (Garber 2009, 149).

Leibniz's response of 24 June 1695 demonstrates his refusal of this complex problem of empirical measure. Emphasizing the *a priori* nature of his measurement of *vis*, and sticking perhaps closer to the Galilean spirit of *a priori* demonstration, Leibniz remarks that any force absorbed by ambient matter is reconstituted in the systematic whole of phenomenal effect. Hence, the participation of ambient matter in the resulting phenomenal effects of motion is constant and thus theoretically (mathematically) independent of the measurement of the invariance of *vis qua* mv^2. Here, Leibniz emphasizes that his use of the Galilean law of falling bodies was only due to its aptness for treating the particular example of the measurement of the *vires* at work across two different pendulums. In fact, if we examine the argument of the *Dynamica*, which was unpublished, we saw that Leibniz already anticipated counter-arguments made on the basis of the contingency of the Galilean law and the motion of the plenum. The universal conservation of mv^2 was based on the immanent action of a body based on the factor of formal effect and state of velocity (GM VI 292).

---

[12] See Aiton 1965, 172.

[13] See Newton 1972, 330–331.

Although a version of the Galilean law of falling bodies certainly played a central role in the formation of Leibniz's theory of *vis*, Leibniz's remarks to Bernoulli here amount to the idea that the central theory of *vires*, which provide the mathematical translation between "entire effect" and "complete cause" through the invariant mv^2, occurs at a distinct level from that of the actual geometrical features and measures of physical motion. The shape and motion of ambient matter is independent of the consideration of the calculation of forces. Bernoulli mistakes the relation between phenomena (the shape, intensity, and change of motion) for the categorically different relation between causes and phenomena. This reinforces the point that I have been making here and indicates the errors that might ensue if the core of the dynamics and its applications are not properly distinguished.

In brief, it appears from Leibniz's treatment of rotational motion *per se* and the elliptical motion of celestial bodies that he saw Newton's gravitational theory as a significant challenge to his dynamics. Despite this challenge, it is also clear that Leibniz saw the Newtonian alternative as a problem resolvable by geometrical means. There are good reasons to disagree with Leibniz here, but we can also appreciate why he thought that these problems did not or, even better, could not rival the foundation that he had developed, through the dynamics, for the *cause* of motion. The reason for this can be traced to the important conceptual difference between the concept of *vis qua* cause and its phenomenal effects already mentioned above. For Newton, the importance of the case of rotational motion is due to its capacity to demonstrate *operational* force as the *cause* of the change of motion. For Leibniz, the cause of motion is structural, and thus operational forces governing the path of motion are secondary to the fundamental problem of *vis* and dynamic causation. This distinction between the core of the dynamics and problems of application concerning the account of actual, say, celestial motion, explains why Leibniz sought merely geometrical explanations for curvilinear motion.

By treating curvilinear motions as complex phenomena reducible to the simple phenomena of linear motions, Leibniz was clearly sidestepping the potentially devastating implication of the absoluteness of rotational acceleration for his long-held position concerning relative motion. Leibniz constrains the expressions of *vis* to those of inertial rectilinear motions. In so doing, the dynamics was immunized from the Newtonian danger. This clarification helps us distinguish between the problem of the cause of motion and the problem of the path that actual motions might take.

With all of this we should not forget that there is no contradiction between Leibnizian *vis* and Newtonian force. Both cohere in a (Newtonian-Classical) physical theory where energy-work and force sit comfortably alongside Galilean relativity and Newtonian absolute space. Nonetheless, for the purposes of understanding Leibniz, the central concept of dynamic cause should be distinguished from the Newtonian concept of force *qua* change of momentum.

6.5   *Vis viva* as Action, Appetition, and Entelechy

The aim of my analysis in the previous section has been to disambiguate the core of Leibniz's systematic dynamics by putting emphasis on the concept of causality that was successively refined from the 1670s to the late 1690s. This is important for making the case that the dynamics relies on a theory of causation across different levels of reality. *Vires* are causes of motion insofar as they cause a group of internally related phenomena subject to Galilean relativity governed by the invariant mv^2. This concept of causation is, again, far from the Newtonian concept of operational cause and thus independent from the strictly "mechanical" change of (the momentum of) corporeal motion. We are thus well on the path of providing an interpretation of the coherence of the dynamics with a theory of autarkic monads. *Vires* govern the phenomena of motion that monads represent subjectively in apperception. Hence, although perception provides the means to determine geometrical features of motion, a theory of *vires* determines the underlying structure of this phenomenon across internally related variations.

The examination of Leibniz's concept of *situs* and its ambient space suggests that, despite the ontological dependence of the existence of space on bodies, the structure of space was absolute and necessary, albeit in a generic or transcendental manner. In this way, we can consider Euclidian-3 space as a necessary condition for the spatialization of bodies in phenomenon. Again here, it would be false to say that bodies are located in space in line with a container-contained relation. We should rather say that bodies "express" a necessary spatial relation by virtue of their existence, despite the fact that this space is itself abstractly distinct from the particular positions of actual bodies.

Now, this relation between bodies and ambient space does not imply anything about motion directly outside of the necessary space that it expresses. Nonetheless, it contributes to an account of how non-spatial *vires* relate to the spatial phenomenon of motion. As we examined earlier, Leibniz's theory of relative motion is the equivalence of hypotheses. Each variation of relative states of rest and motion within a physical system, taken in abstraction, constitutes a different "motion," kinematically speaking. Yet the equivalence across these variations, governed by the invariant *vis*, allows these different motions to constitute a single locomotive event caused by *vis*. Our discussion of space above allows us to grasp a fuller picture of this relationship between *vis qua* cause and phenomena *qua* effect because the gap between cause and effect here is bridged by thinking along the same lines as the bridge between transcendental-generic space and actual *situs*. Just as Leibniz defended his position on space to Clarke through the notion of the "order of situations" instead of just "order" or just "situations," we can think of Leibnizian *vires* as producing a determinate and internally related group of phenomenal variations answering to such an invariant *qua* cause.

We can now solve the puzzle posed at the beginning of this chapter concerning the objective and subjective counterparts of Leibniz's *vires* in the monadic world. A metaphysical world of autarkic monads seemed incapable of accommodating an account of physical reality because any physical event is driven by the monad's internal appetition and represented by the monad's internal perceptions. The fact that these internal appetitions and apperceptions are universally coordinated in pre-established harmony is not, by itself, adequate to providing anything beyond the subjectivism of this account of physical reality.

In our analysis above, however, we see that the objective counterpart of *vires* is part and parcel of, for the lack of a better expression, the "intimate life of the monad." Spatial structure is objective insofar as it is generically absolute and hence part of the generic structure of monadic perception. This "objective counterpart" of space is, of course, not part of metaphysical reality, and no monad is spatially located. Similarly, the objectivity of *vis viva* is a determining *cause* of any locomotive phenomenon regardless of the subjective perception of the phenomenon itself. It is hence a mistake to think that the doctrine of *vis viva* is either subjective and hence inadequate for developing an objective account of corporeal motion, or objective and hence inconsistent with the seemingly exclusively subjective nature of physical reality in the metaphysics of the *Monadology*.

The internal perceptions of coexistent autarkic monads are metaphysically harmonized according to the divine providence of the best of all possible worlds. According to this metaphysical harmony, physical events can only be represented internally to monads which themselves possess no spatial location. However, this internal reality is nonetheless governed by the immanent *vires* of monads. There is a necessary structure for any given spatial perception. Similarly, *vires* are causal in the sense that they produce the necessary structural relations for any given system of locomotive phenomena. Hence, to correlate primitive *vis* to entelechy is not to impute this *vis* to some particular body in mechanical interaction. Indeed, bodies are not monads but rather phenomenal representations of monads.[14] Since any particular motion already implies a range of variations (of rest and motion) for that locomotive event, one could impute this *vis* to any arbitrary body in a system of mechanical interactions under the condition that the invariance of mv^2 is observed. This is universal insofar as the same relation of cause and effect holds whether we consider a local interaction of a system of one- or n-bodies or the parts belonging to the totality of the physical universe. This is simply what it means for mv^2 to be conserved. In turn, this holds for any locomotive phenomenon represented in monadic experience.

The view sketched here gives concrete meaning to Leibniz's correlation of action understood in a strictly metaphysical sense and *actio* understood in the dynamic sense. I shall not enter into the larger metaphysical problems of action understood as "entelechy," which have been sufficiently addressed by many commentators. I only remark that the metaphysical presentation of action, a term that accompanied Leibniz throughout much of his maturity, borrowed from the Thomistic-Scholastic maxim "*actiones [autem]sunt suppositorum*," provides a logical link between the metaphysical activity of substances qua monads and the action of bodies. This metaphysical concept of action, associated with the doctrine of the inclusion of predi-

---

[14]The problem of the "organic" account of bodies in Leibniz's late metaphysics is deliberately bracketed. The point here is simply that monads, which are themselves non-spatial, represent bodies *qua* spatial phenomena. The organic structure of bodies naturally falls under the category of spatial relations.

cates in the subject, develops through the 1670s to the 1690s into the notion of monadic appetition that moves perception from one kinematic frame to the next in a series. In turn, the sum of these individual serial perceptions will constitute the complete life of the autarkic monad. Understood dynamically, however, *vis qua actio* provides the needed account for the physical structure of this very concept of monadic perception and appetition. The causal structure of the physical content of the monadic cinema is determined by *vis qua actio*. Physical reality is thus not reducible to the kinematic succession of perceptions answering to mechanical relations. Rather, motion is governed as phenomenon through a dynamical invariance that stands as the cause of this variational phenomenon.

The metaphysical harmony of monadic-kinematic perceptions of physical reality is thus to be reduced to the dynamics rather than vice-versa. In this understanding, the harmony of the monadic perception of the physical world is *caused* by the internal *vis qua actio* of monads. In physical terms, this harmony is the invariance of *actio* that not only determines the variational perception of particular monads but also the synthesis of these "subjective" perceptions into invariant magnitudes constituted by objective dynamical *vis*. Of course, this is only the case if what we mean by causation between *vis* and extended motion is the relation between non-phenomenal *vis* and phenomenal motion rather than the relation between one empirical phenomenon and another.

6.6   Concluding Remarks

Leibniz's dynamics project was initially aimed at a mere reform of Cartesian mechanics. He eventually saw it more fit to deliver a full critique of mechanics in favor of a "new science" based on the causal nature of *vis*. The project can be understood along its development as Leibniz's own reckoning with the laws of motion and collision and as a new foundation for natural science. However, as Leibniz's metaphysics evolved beyond questions concerning substantial forms of physical bodies, the embrace of autarkic monads in the last decade and a half of Leibniz's work appears to render physical entities as mere phenomena.

This chapter argues that, despite apparent conflict, the dynamics contributes to the development of the monadic metaphysics or is at least consonant with it. The central claim is that the very conception of dynamical causation is a structural one that treats *vis qua* cause as the determination of a group variation of phenomena linked through the invariance of mv^2. Understood dynamically, the autarkic monad "acts" insofar as producing internal representations, phenomena, and their spatial variations that are structurally governed by *actio qua vis viva*. Here, the generic but necessary nature of space helps us understand that, despite the fact that monads are not localized in space, physical relations enjoy a fully determined phenomenal structure. In the same way, although *vires* are not themselves extended, they are the causes of extended phenomena. In turn, dynamic causation provides the crucial link between merely subjective apperceptions of motion in the monad and its objective counterpart, the *vis* inherent in substances *qua* monads.

Bibliography

I. Texts by Gottfried Wilhelm Leibniz

1. Texts Cited by Abbreviation

A:

Leibniz, G.W. 1923. In *Sämtliche Schriften und Briefe*, ed. Academy of Sciences of Berlin. Darmstadt/Leipzig/Berlin: Akademie Verlag.

Cited by Series, Volume, and Page

AG:

Leibniz, G.W. 1989. *Philosophical Essays*. Trans. and ed. R. Ariew and D. Garber. Indianapolis: Hackett Publishing.

C:

Leibniz, G.W. 1903. *Opuscules et fragments inédits*. ed. Louis Couturat. Paris: F. Alcan. Reprinted 1988. Hildesheim: Olms.

GM:

Leibniz, G.W. 1849–1863. *Mathematische Schriften*. 7 vols., ed. C.I. Gerhardt. Berlin/Halle: A. Ascher and H. W. Schmidt.

© Springer International Publishing AG 2017
T. Tho, *Vis Vim Vi: Declinations of Force in Leibniz's Dynamics*, Studies in History and Philosophy of Science 46, DOI 10.1007/978-3-319-59055-4

Cited by Volume and Page

GP:

Leibniz, G.W. 1875–1890. *Die Philosophischen Schriften*. 7 vols., ed. C.I. Gerhardt. Berlin: Weidmannsche Buchhandlung. Reprinted 1960–1961. Hildesheim: Olms.

Cited by Volume and Page

L:

Leibniz, G.W. 1969. *Philosophical Papers and Letters*. 2nd Ed.(trans. Leroy E. Loemker). Dordrecht/Boston: Reidel.

LC:

Leibniz, G.W. 2001. *The Labyrinth of the Continuum: Writings on the Continuum Problem*, 1672–1686. Trans., ed., and intro. R.T.W. Arthur. New Haven/London: Yale University Press.

LdB:

Leibniz, G.W. 2007a. *The Leibniz-Des Bosses correspondence*. Trans., ed., and intro. Brandon C. Look and Donald Rutherford. New Haven/London: Yale University Press.

LdV:

Leibniz, G.W. 2013a. *The Leibniz-De Volder Correspondence. With Selections from the Correspondence Between Leibniz and Johann Bernoulli*. Trans. ed., and intro. Paul Lodge. New Haven/London: Yale University Press.

2. Other Texts of G.W. Leibniz (Not Cited by Abbreviation)

Leibniz, G.W. 1991. Phoranomus seu de potentia et legibus naturae, ed. and annotation by André Robinet. *Physis* 28(2): 429–541 and *Physis* 28(3): 797–885.
Leibniz, G.W. 1994. *Leibniz: La réforme de la dynamique*, ed. Michel Fichant. Paris: Vrin.
Leibniz, G.W. 2004. *Discours de Métaphysique et Monadologie*, ed. and intro. Michel Fichant. Paris: Editions Gallimard.
Leibniz, G.W. 2007b. Phoranomus seu de potentia et legibus naturae, ed. Gianfranco Mormino. In Dialoghi filosofici e scientifici, Gottfried Wilhelm Leibniz, Ed. Francesco Piro, Gianfranco Mormino, and Enrico Pasini. 680–885. Milan: Bompiani.

Leibniz, G.W. 2013b. Principia mechanica. ed. and trans. Richard T.W. Arthur. *The Leibniz Review* 23: 101–105.
Leibniz, G.W., and Samuel Clarke. 2000. *Correspondence*, ed. and intro. Roger Ariew. Indianapolis: Hackett Publishing Company.

II. Texts by Other Authors

Aiton, E.J. 1965. An Imaginary Error in the Celestial Mechanics of Leibniz. *Annals of Science* 21 (3): 169–173.
Aristotle. 2001. *The Basic Works of Aristotle*. Ed. Richard McKeon. New York: Modern Library.
Arthur, Richard T.W. 1994. Space and Relativity in Newton and Leibniz. *British Journal for the Philosophy of Science* 45: 219–240.
Arthur, Richard T.W. 2013. Leibniz's Theory of Space. *Foundations of Science* 18 (3): 499–528.
Barrow, Issac. 1860. *The Mathematical Works of Issac Barrow*. Ed. W. Whewell. Cambridge: Cambridge University Press.
Beeley, Philip. 1996. Kontinuität und Mechanismus. Zur Philosophie des jungen Leibniz in ihrem ideengeschichtlichen Kontext. *Studia Leibnitiana Suppementa* 30: 228–260.
Bernstein, Howard. 1980. Conatus, Hobbes, and the Young Leibniz. *Studies in History and Philosophy of Science* 11: 25–37.
Bernstein, Howard. 1981. Passivity and Inertia in Leibniz's "Dynamics". *Studia Leibnitiana* 13: 97–113.
Bernstein, Howard. 1984. Leibniz and Huygens on the "Relativity" of Motion. *Studia Leibnitiana Sonderheft* 13: 85–102.
Bertoloni Meli, Domenico. 1990. The Relativization of Centrifugal Force. *Isis* 81 (1): 23–43.
Bertoloni Meli, Domenico. 1993. *Equivalence and Priority: Newton Versus Leibniz*. Oxford: Clarendon Press.
Boudri, J. Christian. 2002. *What was Mechanical About Mechanics: The Concept of Force Between Metaphysics and Mechanics from Newton to Lagrange*. Dordrecht: Springer.
Brown, Gregory. 1984. 'Quod ostendendum susceperamus'. What did Leibniz undertake to show in Brevis Demonstratio? *Studia Leibnitiana Sonderheft* 13: 122–137.
Bussotti, Paolo. 2015. *The Complex Itinerary of Leibniz's Planetary Theory*. Basel: Birkhäuser.
Clavius, Christophorus. 1611–1612. *Opera Mathematica Vol. I*. Moguntiae/Mainz. Reprinted in 1999. Hildesheim/New York: Olms-Weidmann.
Costabel, Pierre. 1960. Leibniz et la dynamique: Les textes de 1692. Paris: Hermann. English edition: Costabel, Pierre. 1973. *Leibniz and Dynamics*: the texts of 1692 (trans: R. E. W. Maddison). Ithaca: Cornell University Press.
D'Alembert, Jean. 1785. *Encyclopédie méthodique*.
Darrigol, Olivier. 2014. *Physics and Necessity*. Oxford: Oxford University Press.
De Risi, Vincenzo. 2007. *Geometry and Monadology*. Basel: Birkhäuser.
De Risi, Vincenzo. 2012. Leibniz on Relativity: The Debate Between Hans Reichenbach and Dietrich Mahnke on Leibniz's Theory of Motion and Time. In *New Essays in Leibniz Reception*, ed. Ralf Krömer and Yannik Chin-Drian, 143–185. Basel: Springer.
De Risi, Vincenzo. 2015. Leibniz's Theory of Parallel Lines. In *Mathematizing Space*, ed. Vincenzo de Risi, 1–13. Birkhäuser: Basel.
Duchesneau, François. 1994. *La Dynamique de Leibniz*. Paris: Vrin.
Duchesneau, François. 1998. Leibniz's Theoretical Shift in the Phoranomus and Dynamica de Potentia. *Perspectives on Science* 6 (1–2): 77–109.
Duchesneau, François, 2010. *Leibniz, le Vivant Et L'Organisme*. Paris: Vrin.
Dugas, René. 1988. *A History of Mechanics*. Trans. J.R. Maddox. New York: Dover.

Duhem, Pierre. 1991. *The Origins of Statics*. Trans. Grant F. Leneaux, Victor N. Vagliente, Guy H. Wagener. Boston Studies in the Philosophy of Science Vol. 123. Dordrecht: Kluwer.

Earman, John S. 1989. *World Enough and Space-Time: Absolute versus Relational Theories of Space and Time*. Cambridge, MA: MIT Press.

Erlichson, Herman. 1997. The Young Huygens Solves the Problem of Elastic Collisions. *American Journal of Physics* 65 (2): 149–154.

Eustachio a Sancto Paolo. 1614. *Summa philosophiae quadripartita*. Pars III. Paris: Carolum Chastelain.

Fichant, Michel. 1994. Introduction. In *Leibniz: La réforme de la dynamique*, ed. Michel Fichant, 9–65. Paris: Vrin.

Fichant, Michel. 1995. De la puissance à l'action: la singularité de la Dynamique. *Revue de Métaphysique et de Morale* 100 (1): 49–81.

Fichant, Michel. 2004. Introduction. In *G.W. Leibniz: Discours de Métaphysique et Monadologie*, ed. Michel Fichant, 7–140. Paris: Editions Gallimard.

Fichant, Michel. 2016. Les dualités de la dynamique leibnizienne. *Lexicon Philosophicum* 4: 11–41.

Gabbey, Alan. 1971. Force and Inertia in Seventeenth-Century Dynamics. *Studies in History and Philosophy of Science* 2 (1): 1–67.

Gale, George. 1984. Leibniz' Force: Where Physics and Metaphysics Collide. In *Leibniz' Dynamica*, ed. A. Heinekamp, 62–70. Stuttgart: Steiner.

Gale, George. 1988. The Concept of 'force' and Its Role in the Genesis of Leibniz' Dynamical Viewpoint. *Journal of the History of Philosophy* 26 (1): 45–67.

Galilei, Galileo. 1898. *Le Opere di Galileo Galilei Vol. VIII*. Firenze: G. Barbera.

Galilei, Galileo. 1960. *On Motion and On Mechanics*. Trans., intro. and notes I.E. Drabkin and Stillman Drake. Madison: University of Wisconsin Press.

Galilei, Galileo. 1974. *Two New Sciences*. Trans. and ed. Stillman Drake. Madison: University of Wisconsin Press.

Garber, Daniel. 1985. Leibniz and the Foundations of Physics: The Middle Years. In *The Natural Philosophy of Leibniz*, ed. K. Okruhlik and J.R. Brown, 27–130. Dordrecht: D. Reidel Publishing Company.

Garber, Daniel. 2009. *Leibniz: Body, Substance, Monad*. Oxford/New York: Oxford University Press.

Gueroult, Martial. 1934. *Dynamique et métaphysique Leibniziennes*. Paris: Les Belles Lettres.

Hankins, Thomas. 1990. *Jean D'Alembert: Science and the Enlightenment*. New York: Gordon and Breach.

Hermann, Jakob. 1716. *Phoronomia, sive de viribus et motibus corporum*. Amstelaedami (Amsterdam): R. & G. Wetstenios H.FF.

Hobbes, Thomas. 1999. In De Corpore: Elementorum philosophiae sectio prima, ed. Karl Schuhmann. Paris: J. Vrin.

Huygens, Christiaan. 1669. Extrait d'une lettre de M. Huygens. *Journal des sçavans* 1672: 22–24.

Huygens, Christiaan. 1929. *Oeuvres Completes,* Vol. 16. La Haye: Martinus Hijhof.

Huygens, Christiaan. 1977. The motion of colliding bodies (trans. Richard J. Blackwell). *Isis* 68 (4): 574–597.

Huygens, Christiaan. 1993. 'Penetralia motus': la fondazione relativistica della meccanica in Christiaan Huygens, con l'edizione del Codex Hugeniorum 7A. Ed. Gianfranco Mormino. Firenze: La Nuova Italia.

Iltis, Carolyn. 1971. Leibniz and the Vis Viva Controversy. *Isis* 62 (1): 21–35.

Itokazu, Anastasia Guidi. 2009. On the Equivalence of Hypotheses in Part I of Johannes Kepler's New Astronomy. *Journal for the History of Astronomy* 40 (2): 173–190.

Jauernig, Anja. 2008. Leibniz on Motion and the Equivalence of Hypotheses. *The Leibniz Review* 18: 1–40.

Jauernig, Anja. 2009. Leibniz on Motion- Reply to Edward Slowik. *The Leibniz Review* 19: 139–147.

Jungius, Joachim. 1627. *Geometria empirica*. Rostock: Haeredum Richelianorum.

Jungius, Joachim. 1638. *Logica hamburgensis*. Hamburg: Bartholdi Offerman.

Jungius, Joachim. 1699. Phoronomica. In *Opuscula Mathematica*, ed. Johannis Aldolfi Tassi, Henricum Siverum, and Balthasar Mentzer, 1–46. Hamburg: Gottfried Liebezeit.

Kepler, Johannis. 1984. *The Birth of History and Philosophy of Science: Kepler's a Defence of Tycho Against Ursus with Essays on Its Provenance and Significance*. Ed. and anno. with supplementary essays by N. Jardine. Cambridge: University of Cambridge Press.

Kepler, Johannes. 1990. *Gesammelte Werke* Vol. III. Ed. Kepler-Kommission Der Bayerischen Akademie der Wissenschaften. München: C.H. Beck'sche Verlagsbuchhandlung.

Kepler, Johannes. 1992. *New Astronomy*. Trans. William H. Donahue. Cambridge: Cambridge University Press.

Knobloch, Eberhard. 2008. Generality and Infinitely Small Quantities in Leibniz's Mathematics – The Case of His Arithmetical Quadrature of Conic Sections and Related Curves. In *Infinitesimal Differences. Controversies Between Leibniz and His Contemporaries*, ed. E. Goldenbaum and D. Jesseph, 172–183. Berlin: De Gruyter.

Leibniz, G. W. 1981. *New Essays on Human Understanding*, ed. and trans. by Peter Remnant and Jonathan Bennett. Cambridge: Cambridge University Press.

Lodge, Paul. 1997. Force and the Nature of Body in Discourse on Metaphysics §§17–18. *Leibniz Society Review* 7: 116–124.

Lucretius (Carus), Titus. 2002. *On the Nature of Things*. Trans. by W.H.D. Rouse and rev. by Martin F. Smith. Cambridge, MA: Loeb Classical Library.

Mariotte, Edme. 1673. *Traitté de la percussion ou les chocq des corps*. Paris: Chez Estienne Michallet.

McDonough, Jeffrey K. 2010. Leibniz's Optics and Contingency in Nature. *Perspectives on Science* 18 (4): 432–455.

Mercer, Christia. 2007. *Leibniz's Metaphysics: Its Origins and Development*. Cambridge: Cambridge University Press.

Mormino, Gianfranco. 1996. The Philosophical Foundations of Huygens's Atomism. In *De zeventiende eeuw, Vol. 12*, ed. L. Palm, 74–82. Hilversum: Verloren.

Mormino, Gianfranco. 2011. Leibniz entre Huygens et Newton: force centrifuge et relativite du mouvement dans les lettres de 1694. In *Natur und Subjekt. IX. Internationaler Leibniz-Kongress Vorträge Vol. 2*, ed. Herbert Breger, Jurgen Herbst, and Sven Erdner, 697–705. Hannover: Druckerei Hartmann.

Newton, Issac. 1972. *Mathematical Principles of Natural Philosophy,* Vol. I. Trans. Andrew Motte and ed. Florian Cajori. Berkeley/Los Angeles/London: University of California Press.

Papin, Denis. 1689. De gravitatis causa et proprietatibus observationes. *Acta Eruditorum*: 183–189.

Proclus, Diadochus. 1970. *A Commentary on the First Book of Euclid's Elements*. Trans. Glenn R. Morrow. Princeton University Press.

Ranea, Alberto Guillermo. 1989. The a priori Method and the Actio Concept Revised. *Studia Leibnitiana* 21 (1): 42–68.

Reichenbach, Hans. 1928. *Philosophie der Raum-Zeit-Lehre*. Berlin: De Gruyter. English edition: Reichenbach, Hans. 1957. *The Philosophy of Space and Time* (trans: Maria Reichenbach and John Freund). New York: Dover Publications.

Roberts, John T. 2003. Leibniz on Force and Absolute Motion. *Philosophy of Science* 70: 553–573.

Roberval, Gilles P. 1996. *Eléments de géometrie de G.P. Roberval*. Ed. Vincent Jullien. Paris: Vrin.

Robinet, André. 1984. Dynamique et fondements métaphysiques. *Studia Leibnitiana Sonderheft* 13: 1–25.

Robinet, André. 1988. *G.W. Leibniz. Iter Italicum*. Florence: Leo Olschki.

Robinet, André. 1989. Les surprises du phoranomus: L'art d'inventer, le principe d'action, et la dynamique. *Les Etudes philosophiques* 2: 171 186.

Sleigh, Robert Jr. 1990. *Leibniz and Arnauld: A Commentary on Their Correspondence*. New Haven: Yale University Press.

Slowik, Edward. 2006. The 'Dynamics' of Leibnizian Relationism: Reference Frames and Force in Leibniz's Plenum. *Studies in History and Philosophy of Modern Physics* 37: 617–634.

Stammel, Hans. 1984. Der Status der Bewegungsgesetze in Leibniz' Philosophie und die apriorische Methode der Kraftmessung. In *Leibniz' Dynamica*, ed. A. Heinekamp, 180–188. Stuttgart: Steiner.

Stan, Marius. 2016a. Huygens on Inertial Structure and Relativity. *Philosophy of Science* 83 (2): 277–298.

Stan, Marius. 2016b. Perpetuum mobiles and eternity. In *Eternity: The History of a Concept*, ed. Yitzhak Melamed, 173–178. Oxford: Oxford University Press.

Stein, Howard. 1977. Some Philosophical Prehistory of General Relativity. In *Foundations of Space-Time Theories*. Ed. John Earman et. al. 3–49. Minneapolis: University of Minnesota Press.

Strum, Leonhard Christoph. 1706. *Natura et constitutione matheseos*. Frankfurt: Jer. Schrey and Joh. Christ. Hartmann.

Spector, Marshall. 1975. Leibniz vs. the Cartesians on Motion and Force. *Studia Leibnitiana* 7 (1): 135–144.

Suárez, Francisco. 2015. *Disputationes Metaphysicae*. http://homepage.ruhr-uni-bochum.de/Michael.Renemann/suarez/index.html. Accessed 28 Mar 2016.

Szabó, Istvan. 1979. *Geschichte der mechanischen Prinzipen*. 2nd ed. Basel: Birkhäuser.

Torricelli, Evangelista. 1644. *Opera Geometrica*. Florence: Amadoro Massa and Lorenzo de Landis.

Voelkel, James. 2001. *The Composition of Kepler's Astronomia Nova*. Princeton: Princeton University Press.

Wallis, John. 1670. *De Mechanica: sive, De motu, tractatus geometricus*. London: Gulielmi Godbit.

Wallis, John. 1695. *Opera Mathematica Vol. I*. Oxford: E. Theatro Sheldoniano.

Watkins, Eric. 1998. From Pre-established Harmony to Physical Influx: Leibniz's Reception in Eighteenth Century Germany. *Perspectives on Science* 6 (1): 136–203.

Weigel, Erhard. 1776. *Idea totius: Encyclopaediae mathematico-philosophica*. Jena: Johannem Meyerum.

Zedler, Johann Heinrich. 1731–1754. *Grosses vollständiges Universal-Lexicon aller Wissenschafften und Künste*. Halle/Leipzig: Johann Heinrich Zedler.

Index

© Springer International Publishing AG 2017   145
T. Tho, *Vis Vim Vi: Declinations of Force in Leibniz's Dynamics*, Studies in
History and Philosophy of Science 46, DOI 10.1007/978-3-319-59055-4

The manufacturer's authorised representative in the EU is Springer
Nature Customer Service Centre GmbH, Europaplatz 3, 69115 Heidelberg,
Germany. If you have any concerns regarding our products, please
contact ProductSafety@springernature.com

Printed and bound by CPI Group (UK) Ltd, Croydon, CR0 4YY

27/04/2026

02097572-0013